THE CROSS-STAFF

Good luck
on a
good course

Frank Hunter

The Figure of the Crofs-Staff.

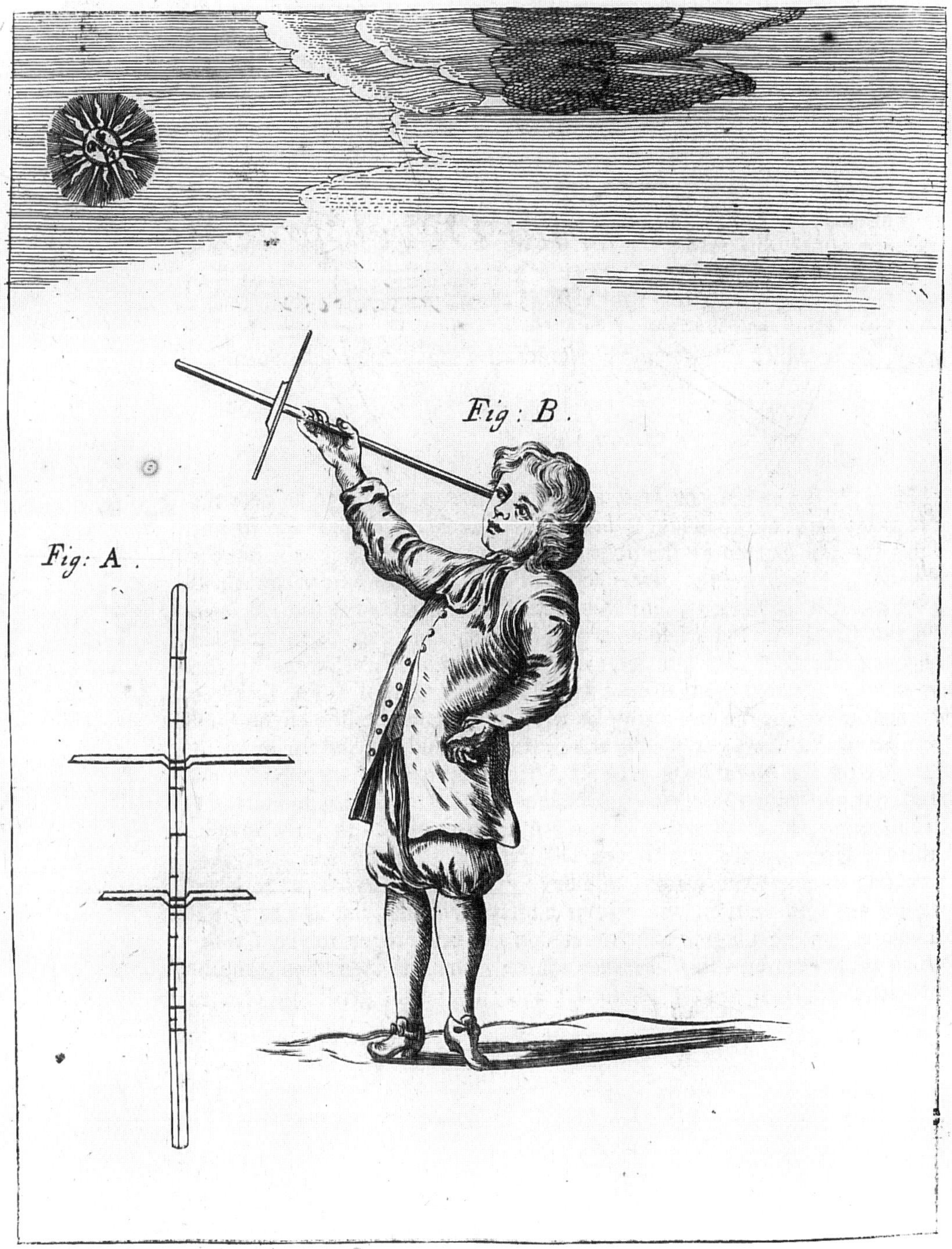

THE CROSS-STAFF

History and Development of a Navigational Instrument

BY

W.F.J. Mörzer Bruyns *FRIN*

VEREENIGING NEDERLANDSCH HISTORISCH SCHEEPVAART MUSEUM
RIJKSMUSEUM NEDERLANDS SCHEEPVAARTMUSEUM AMSTERDAM
WALBURG INSTITUUT

DRAWINGS
Kurt de Reus, Zoelmond

The publication of this book was generously supported by the following foundations:
Stichting Dr Hendrik Muller's Vaderlandsch Fonds, Amsterdam
Oostersche Handel en Reederijen, Amsterdam
J.E. Jurriaanse Stichting, Rotterdam
M.A.O.C. Gravin van Bylandt Stichting, 's-Gravenhage
J.C. Ruigrok Stichting, Haarlem
Stichting K.F. Hein Fonds, Utrecht

Cover: 'The Element Air', with a cross-staff, a celestial globe, a quadrant and a mariner's astrolabe.
Oil on panel by Nicolaes de Heldt Stockade, *c.* 1662. Ceiling decoration in the large hall of the Trippenhuis in Amsterdam.
Courtesy, Koninklijke Nederlandse Akademie van Wetenschappen, Amsterdam. Photograph, Gert Jan van Rooij, Amsterdam.
Frontispice: 'The figure of the cross-staff.'
From Jonas Moore, *A New Systeme of Mathematics*, London, 1681.
Nederlands Scheepvaartmuseum, Amsterdam.

Printed by: Walburg Druk, Zutphen

CIP-ISBN 906011.907.X

For my parents

'I Present vnto thee (gentle Reader) in one small Volume, the vse of two Mathematicall Instruments, the one is the Crosse Staffe: the other, the Iacobs Staffe.'

Thomas Hood 1590

Contents

Foreword

The internationally important collection of the *Vereeniging Nederlandsch Historisch Scheepvaart Museum*, was founded in 1916. Since 1973 it has been housed in the seventeenth-century building of the *Nederlands Scheepvaartmuseum*, Amsterdam. Numerous objects relating to our maritime past are preserved, in order to present to a broad public the history, culture, and art that are connected with the sea.

The technique of navigation is, of course, one of the fundamental sciences for those who roam the seas. For many centuries seamen have used the stars with the help of navigational instruments. One of these instruments was the cross-staff, seemingly a quite insignificant object, consisting of a straight piece of wood with vanes which were often lost when the instrument was no longer in use. It has no glamour at all. As is the destiny of many such objects, the cross-staff became a rarity. Once many thousands of these instruments were used by seamen all over the world. Now no more than 95 are known to have survived. This book aims to sum up all the recent knowledge on the cross-staff. The author, the Museum's curator of navigation, has, for the past twenty-five years, acquired globes, charts and instruments, including cross-staffs, adding these to the collection of the *Vereeniging*. He also sought out cross-staffs elsewhere in the world. In this book, Mörzer Bruyns connects archival and printed sources, iconographical information and, last but not least, the objects themselves. Thus he has put into practice an essential feature of our research policy in the Museum, that is, to integrate objects from the collections with other information in order to formulate new insights.

It is our firm belief that this publication of the *Vereeniging Nederlandsch Historisch Scheepvaart Museum* and the *Nederlands Scheepvaartmuseum* will provide for the years to come a sound basis for all who undertake research into one of the most popular navigational instruments in history: the cross-staff.

A.N. van der Wiel
Chairman

Drs. J. Bakker
Director

Preface

This book consists of two equally important parts. The first is an introduction to the history and use of the instrument, the second a checklist, with descriptions, of ninety-five cross-staffs, in collections throughout the world.

The introductory section starts with a survey of the history of the art of astronomical navigation and navigational instruments, between *c.* 1500 and *c.* 1800, the period during which the cross-staff was used at sea. The thread in this chapter, which is based on modern published research, is the development of astronomical navigation, focusing on instruments for the measurement of altitude.

In the chapter, Historiography and Sources on the Cross-Staff, the instrument's history as published to date is evaluated. Books and manuscripts contemporary to the instrument itself are an important and rich source. These consist of navigational manuals and sea-atlases written by scholars and seamen, and of archival material, mainly of the Dutch East India Company, the VOC. Also, iconographical material as a provider of information so far not used for this type of instrument is discussed. Lastly, the surviving instruments themselves as a source of knowledge are introduced.

In the third chapter the cross-staff, with its variety of names, is introduced, defined, and described. Instruments that were derived from the cross-staff are discussed, and this chapter contains a description of how the cross-staff was used, and the errors to which it was subject. Finally, the different components, such as the vanes and aperture disk, and the methods for constructing the scales, are dealt with. This chapter is based on published modern research, except for the construction of the scales, for which printed sources contemporary to the cross-staff were used.

Makers, sales, and the economic value of the instrument, and the seamen who used it, are dealt with in the chapter on the Production and Diffusion of the Cross-Staff. Here, focus is on the VOC, for which, unlike most historical maritime organizations, archival material is abundantly available. Some of the information has been placed in the second and third appendices.

In the final and main chapter of the introductory section, the development of the surviving staffs (including the scales) and vanes in the Checklist, is described, and compared with contemporary written sources.

The Checklist, a chronological catalogue of cross-staffs in collections, is the result of eighteen years of tracing and studying these instruments. Out of the total of ninety-five, four are lost or could not be retrieved. Of the remaining, eighty have been seen by the author. Sixty-five cross-staffs were made in the Netherlands, while an additional eleven very probably originate from that country. The rest were made in England, France, North America, Spain, Germany or Scandinavia.

The first appendix contains notes on makers, with the emphasis on the Dutch and British. This bias results from the readily available information in the Amsterdam municipal archives, and from Project SIMON, and the lack of such sources elsewhere. In the last four appendices, some of the data in the Checklist are listed to make comparison and future identification easier.

Acknowledgements

Soon after being made responsible for the collection of navigational instruments in the Scheepvaartmuseum, I became intrigued by cross-staffs, of which the Museum owns several. The graduated staff, made of black wood, with vanes, was pre-eminently part of a seaman's equipment. One finds them illustrated in decorative cartouches of sea-charts and title pages of sea-atlases, but also in allegorical pictures and still-life paintings, together with symbols of time and the mortality of man, such as hour-glasses and skulls. Finally, as a symbol of their profession, merchant captains and admirals were often portrayed with a cross-staff.

When, in 1976, I started collecting data on cross-staffs surviving in museums and private collections, it became apparent that relatively few remained. This book is the result of the continued search and I sincerely hope that it will be of help and guidance to all those who share my interest in historical navigational instruments.

I am grateful to the Editorial Committee, the members of which, from the very beginning, showed great interest in the project. All along they gave me guidance and valuable advice.

I want to thank Dr G.C. Clifton of Project SIMON, at the National Maritime Museum, London, for supplying the names of British cross-staff makers, and Mrs R.V. Wallis of the Project for Historical Bibliography in Newcastle-upon-Tyne, for information on book titles. I am indebted to Mr Peter von Busch, former Director of the Marinmuseum in Karlskrona, and to Mrs Elvira Mata i Enrich, Director of the Museu Maritim de Barcelona, for, respectively, translating parts of a Swedish and a Spanish textbook into English, the contents of which would otherwise have been inaccessible to me. I should like to thank Mr G.J.D. Wildeman, acting librarian of the Nederlands Scheepvaartmuseum, for patiently helping in the reading room, and Dr R.B. Prud'homme van Reine for his assistance with the illustrations. I should also like to thank my wife, Sybrich, for encouraging and supporting me, especially in the final months during which the manuscript was completed. In addition, she spent many hours at the Amsterdam Gemeentearchief tracing names of cross-staff makers. Finally, I thank Mrs Helen Turner for reading the whole manuscript and for patiently correcting my English.

Without the help of many individuals in various museums throughout the world it would not have been possible to complete this book. Although, over the years, I have had the opportunity to see most of the existing staffs, I still had to rely on the help of colleagues for checking storerooms to see whether staffs, sometimes known from literature only, were still present. Some, at my request, made rubbings or drawings of the stamps, and provided photographs. Also, in my search for cross-staffs, colleagues informed me when they came across yet another example unknown to me. For their help I am indebted to the following persons:

In the Netherlands to Drs J. Braat, Curator Rijksmuseum, Amsterdam; Dr P.C.J. van der Krogt, Rijksuniversiteit, Utrecht; Mr Th. Fruithof, Curator, Zuiderzeemuseum Enkhuizen; Mr T. Helperi Kimm, formerly Director of the Noordelijk Scheepvaartmuseum, and Drs Joke Schuller tot Peursum-Meijer, formerly Curator, Universiteitsmuseum, both in Groningen; Dr P. Wisse, Curator, Museon,

The Hague; Drs H.W. Saaltink, formerly Curator, Westfries Museum, Hoorn; Drs E. Kramer, Head of Collections, Fries Museum, and Mr J. Zijlstra, both in Leeuwarden; Mrs Els M. Jacobs MA, Leiden; Drs L.M. Akveld, Curator Maritiem Museum 'Prins Hendrik', Rotterdam; Mr S. ten Hoeve, Director, Fries Scheepvaart Museum, Sneek; Dr J.C. Deiman, Curator, Universiteitsmuseum Utrecht; Dr J.P. van de Voort, formerly of the Visserijmuseum, Vlaardingen; Mr A. Flonk and Mr G. de Weerd, former and present Curators, Gemeentelijk Museum 't Behouden Huys, West-Terschelling. In the United Kingdom to Dr Allen D.C. Simpson, Curator, National Museums of Scotland, Edinburgh; Mr Gerald M. Forty; Mr J.C. Robinson, formerly Curator, and Mr G.P. FitzGerald, Assistant Curator, Science Museum; Mr Alan N. Stimson and Dr Kristen Lipincott, Curators, National Maritime Museum, Greenwich and Mrs Molly Freeman, all in London; to Dr Kate Harris, Librarian and Archivist to the Marquess of Bath, Longleat House, Warminster, Wiltshire. In Germany to Dr Uwe Schnall, Curator, Deutsches Schiffahrtsmuseum Bremerhaven; Mr Wolfgang Köberer, Frankfurt a. M.; Dr Karin Eckhardt, Museum für Kunst und Gewerbe; Kapitän Manfred Kuleisa, retired Regierungsdirektor, Deutsches Hydrographisches Institut; Dr Boye Meyer-Friese, Curator, Altonaer Museum, and Dr Carsten Prange, Curator Museum für Hamburgische Geschichte, all in Hamburg; Miss Julie N. Seddon Jones MA, formerly of the Deutsches Museum, Munich; Dr Claudia Lichte, Würtembergisches Landesmuseum Stuttgart; Mr Werner Ganzel, Dr.-Carl-Haeberlin-Friesen-Museum, Wyk auf Föhr. In the United States of America to Mr & Mrs Roderick S. Webster, The Adler Planetarium, Chicago, Illinois; Mr Donald Treworgy, Mystic Seaport Museum, Mystic, Connecticut; Mr William Andrewes, formerly Curator The Time Museum Rockford, Illinois; Mr Philip C.F. Smith, Dr Paul F. Johnston, former Curators, and Dr Daniel R. Finamore, present Curator of Maritime History, Peabody & Essex Museum, Salem, Massachusetts. In Belgium to Mme J.L. Dosogne, formerly Curator, Musées Royaux d'Art et d'Histoire; Professor P. Melchior, formerly of l'Observatoire Royal, and Mr Jan De Graeve, all in Brussels. In France to Rear-Admiral Francois Bellec, Director, Musée de la Marine, and to Dr Daniëlle Fauque, both in Paris. In Canada to Mr Leonard G. McCann, formerly Curator, Vancouver Maritime Museum, Vancouver, British Columbia. In Sweden to Mr Manne Dunge, formerly Curator, Sjöfartsmuseet i Göteborg, and currently Curator, Marinmuseum, Karlskrona; Dr Arne Losman, Curator, Skoklosters slott, Skokloster; Dr Sibylla Haasum, Director, and Mr Bert Westenberg, Senior Curator, Statens Sjöhistoriska Museum, Stockholm. In Norway to Mr Bård Kolltveit, Director, Norsk Sjøfartsmuseum, Oslo and to Mrs Eli Rasmussen, Bergens Sjøfartsmuseum, Bergen. In Denmark to Captain Søren Thirslund, and to Mr Hans Jeppesen, Director, Handels- og Søfartsmuseet på Kronborg, Helsingør; Mr Peter Kristiansen, Rosenborg Castle, Copenhagen. In Portugal to Captain A. Estacio dos Reis, formerly Curator, Museu de Marinha, Lisbon. In Finland to Mr Risto Halme(†) and to Dr Leena Sammallahti, former and present Head, Section for Maritime History, Museovirasto, Helsinki. In Spain to Dr Maria-Pilar de San Pio, formerly Curator, Museo Naval de Madrid. In Italy to Dr Paolo Brenni, Istituto e Museo di Storia della Scienza, Florence.

Amsterdam, July 1994

W.F.J. Mörzer Bruyns

An Introduction to the History of Astronomical Navigation and Navigational Instruments, 1500 - 1800

In the fifteenth century, navigators from the Iberian peninsula were the first West-Europeans to sail beyond local waters, and in doing so laid the foundation of astronomical navigation.[1] They developed techniques and methods necessary to navigate out of sight of the coastline. Using the magnetic compass, probably introduced at sea in the twelfth century, the Portuguese, in 1481, were the first to cross the Equator, and in 1487 rounded the Cape of Good Hope. America was reached in 1492 and the Malabar coast of India in 1498. Portuguese and Spanish navigators, in the fifteenth century, were capable of calculating their latitude. They could not, however, find their longitude.[2]

In 1514 Johann Werner of Nurnberg, in his comment on Ptolemaeus' *Geographia*, published a proposal to find the longitude by measuring the distance between the moon and a fixed star.[3] In 1524 Peter Apian, in *Cosmographicus Liber*, elaborated on Werner's suggestion and provided an illustration of that method. In 1530 the Louvain scholar, Gemma Frisius, in *De Principiis Astronomiae & Cosmographiae* proposed that the difference in longitude between two places could be measured with a clock. The instrument for measuring the distance between the moon and a star, the tables needed for that method, and the extremely accurate time-keeper necessary for Frisius' method, were not developed with success until the mid-eighteenth century.[4] This, however, did not stop scholars, scientists, and inventors from searching for methods to solve the problem. They were stimulated by rewards in money awarded by the Spanish Kings Philip II in 1567, and Philip III in 1598, by the States of Holland and the States General of the Dutch Republic around 1600, and by the British Parliament in 1714.[5]

Sixteenth-century seamen were also not capable of accurately ascertaining their speed passing through the water. Until the end of the century, sailed distances had to be estimated, thus depending greatly on the experience of the seamen.[6] The combination of the estimated distance run and the observed latitude gave a position.

As the world was explored and charted in a North-South direction, an increasing number of latitudes of coastal places was established. Subsequently, these were used for 'sailing down the latitude'.[7] To cross the Atlantic

Measuring the lunar distance with a cross-staff. From P. Apianus, *Cosmographie*, Antwerp, 1553. Nederlands Scheepvaartmuseum, Amsterdam.

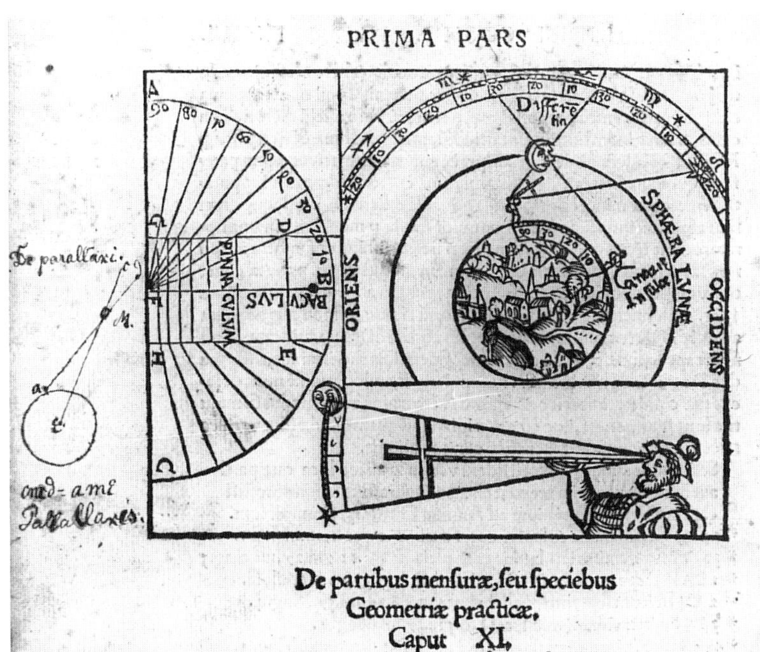

Ocean with this method, a ship would sail along the coast of Europe until reaching the latitude of the place of destination. She would then be sailed due West, checking the latitude daily and, if necessary, correcting the position by sailing West-by-North or West-by-South. The latitude was found by measuring the sun's altitude at noon, using a nautical quadrant, a mariner's astrolabe, or a cross-staff. At night, in the Northern hemisphere, using the same instruments, the Pole star could be observed for the same purpose. The earliest known use of a quadrant by a mariner, in fact a simplified astronomer's instrument, dates from 1460.[8] The mariner's astrolabe, like the quadrant originally an astronomer's instrument, was also introduced at sea in the fifteenth century. The earliest mention dates from 1481, during a voyage by Portuguese down the West coast of Africa.[9] In order to convert the sun's altitude into the observer's latitude it was necessary to know the sun's declination.[10] Around 1509 this declination, which varies with the date, was available in print for Portuguese navigators in a book titled *Regimento do Estrolabio e do Quadrante*.[11] Knowing the altitude and the declination of the observed celestial body, a mariner using the algebraical formula 'latitude equals declination plus zenith distance', could calculate his latitude.[12] Part of the calculation could be avoided if the instrument used for the observation also had a scale for the altitude's complement, the zenith distance. The cross-staff was also introduced at sea by the Portuguese, but not until around 1515. Apparently, sixteenth-century Spanish and Portuguese seamen used it specifically to measure the altitude of the Pole star. For the sun's altitude, they preferred the mariner's astrolabe and the nautical quadrant. Finding latitude from the altitude of the Pole star was relatively easy, as the Pole star is situated almost directly above the North Pole, making its altitude almost equivalent to the observer's latitude. A correction named 'regiment for the North Star', amounting to a few degrees, had to be applied to the observation in order to correct it for the distance between Pole star

Observing the Pole star. The earliest illustration showing a mariner's cross-staff in use. From P. de Medina, *Regimento de Navigacion*, Sevilla, 1552. Nederlands Scheepvaartmuseum, Amsterdam.

and celestial North Pole.[13] As navigation by Europeans in the Southern hemisphere became more frequent, the Southern Cross was sometimes used for latitude observations in the same way as the Pole star, although here a far larger correction, the 'Regiment of the Southern Star', had to be applied.[14]

In order to control trade and navigation between Spain and the Indies, the Spanish monarchs established the 'Casa de Contratación' at Seville, headed by the 'Piloto Mayor'. Initially these functionaries were accomplished navigators, such as Amerigo Vepucci, but in later years they were scholars. Pilots, cosmographers, and cartographers were employed in order to systematize instructions, examinations and certification of masters. Furthermore, at the Casa, a master chart of all known seas was maintained and navigational instruments were examined. Not only maintaining safety and efficiency, but also the secrecy of navigational and cartographical information was the duty of the Casa. This was severely infringed when, in 1548, the Piloto Mayor, Sebastian Cabot, absconded to England with all the secrets of the Casa.[15] It was the beginning of a flow of knowledge of navigational methods and techniques, and of cartography, to North-European countries. Soon navigational manuals were translated. The Spanish *Arte de Navegar* (Valadolid, 1545) by Pedro de Medina was published in French in 1553, in Dutch in 1580, and in English in 1581. Martin Cortes' *Breve Compendio de la Sphera y de la Arte de Navegar* (Seville, 1551), was the basis for William Bourne's *A Regiment for the Sea* (London, 1574). As these manuals became available in North-European countries, voyages in search of sea routes to the wealth of the Orient were undertaken.

In 1595 the back-staff, later to be known as the Davis quadrant, was introduced. It was based on the cross-staff, but, as the name implies, used with the observer's back towards the sun. Over the next two centuries the Davis back-staff was a popular instrument among seamen.

One of the organizations which benefitted from the spread of navigational knowledge was the Dutch East India Company (VOC), a company consisting of six Chambers, in fact small companies.[16] It was founded in 1602, following a number of commercially succesful voyages between the Dutch Republic and the Indonesian Archipelago, in the period 1595-1602. The masters and mates on some of these early Dutch voyages had received tuition from the Amsterdam cartographer, Petrus Plancius. He had managed to acquire charts from the Spanish Piloto Mayor, Barthelomeo de Laso, and combined these with information gathered by Dutchmen, who had served under the Portuguese in Asia.[17] The VOC appointed Plancius as its official cartographer, making him responsible for preparing the charts necessary for the voyages to Asia. Notable successors in this office were three generations of the Blaeu family, Isaac de Graaf and, after 1743, the Van Keulen family, until the VOC was dissolved in 1799.[18] In 1619, a private teacher in the art of navigation in Amsterdam, Cornelis Jansz. Lastman, was appointed the first examiner of the mates of the Amsterdam Chamber.[19]

In 1654, the VOC instructed the examiner and the cartographer to compile a list of navigational equipment to be supplied to each ship sailing to Asia. Besides charts, instruments and sea-atlases, the list, which was first printed in 1655, also contained the titles of navigational manuals and declination tables. The VOC ruled that this equipment would have to be accounted for by masters and mates after each

Detail from 'Allegory of Sight', showing an early seventeenth century cross-staff. Oil on panel by Jan Breughel I and Peter Paul Rubens, 1617. Courtesy, Museo del Prado, Madrid.

voyage. This measure was inspired by the fear of cartographical information falling into the hands of rivals in Asia. Over the years the list was updated according to developments in cartography, the art of navigation, and instrument making, and were subsequently reprinted. Major adjustments and reprints date from 1675, 1731 and 1747.[20] As far as astronomical navigation is concerned, the 1655 list shows that, in addition to cross-staffs, ships were equipped with mariner's astrolabes, back-staffs and a nautical almanac for the declination. After 1670, the provision of astrolabes was terminated. They were expensive, not very practical in use, and their graduation was less accurate than that of the other two instruments. After 1731, the back-staff was no longer provided. For the next sixteen years the cross-staff, which by then had been developed to a high level of accuracy, was the only instrument for altitude measurement on the ships of the VOC.[21]

In 1731 the Englishman, John Hadley, invented a reflecting quadrant for measuring the altitude of sun and stars above the horizon. It had two mirrors and a moveable index arm. The octant, as Hadley's quadrant was named, was easy to handle and gave an accurate reading. Soon it was being used at sea, initially on a limited scale.[22] Although it was used on some VOC ships as early as 1738, it was not until 1747 that the octant was standard issue to VOC ships, for use by the master.[23] Although, with the octant, seamen had obtained an instrument for accurately measuring the altitude of the sun, that observation itself was still restricted to one moment of the day, noon. If, at that time, the sun was obscured by clouds or the horizon by a haze, the observation could not be performed and uncertainty about the observer's latitude was prolonged for at least another twenty-four hours. Stars could be observed for the latitude at other hours of the day, provided their declination was known. But, at night, the horizon is difficult to see, making the outcome of the observation uncertain.

The formal introduction of the octant on ships of the VOC more or less coincided with the availability of a new method of navigation. It was developed by the Dutch teacher in the art of navigation, and examiner of mates and lieutenants in Amsterdam, Cornelis Douwes.[24] In 1747 Douwes introduced a method for finding the latitude by double altitudes, one in the forenoon and one in the afternoon, providing the time between the observations was known. For the first time navigators were no longer solely dependent on the noon observation. Douwes' method was used by seamen in England, France, Germany, Spain, Italy, Sweden, Denmark and the United States.[25]

By the middle of the eighteenth century, the solution of how to find the longitude at sea had come nearer. It had been stimulated by the aforementioned Act of the British Parliament, by which a reward of £ 20.000 was offered for a practical method of finding longitude at sea, within certain limits of accuracy. The methods suggested by Werner and Apian, and by Frisius, became practically available

Keystone from a house in the Westerstraat in Amsterdam, 1638. The house, now demolished, of which one side faced the *Graadboogsgang* [Cross-Staff Alley], was once the home of Cornelis Jansz. Lastman, teacher of navigation and examiner of mates of the Amsterdam Chamber of the Dutch East India Company. Nederlands Scheepvaartmuseum, Amsterdam.

for seamen in the second half of the eighteenth century. In England, John Harrison worked for many years on perfecting timekeepers, and eventually, in 1774 he received the full award for his fourth clock. While Harrison was working in England, in France Ferdinand Berthoud and Pierre le Roy were also developing marine timekeepers. The result of all these efforts was that, in the last two decades of the century, ships' chronometers were being commercially produced and were available for finding the longitude at sea.
Some twenty-five years before, the German astronomer, Tobias Mayer, had calculated tables for the method of lunar distance. They provided the distance, in angles of arc, between the moon and a number of fixed stars, including the sun, for a period in the future and for a standard meridian. When using Mayer's tables and an octant to find his longitude by lunar distance, in 1769, the English naval officer, John Campbell, found that the angle between the moon and a star is often larger than 90°, the maximum range of an octant. He therefore instructed the London instrument maker, John Bird, to make an octant with an angle extended to 120°, thus creating a sextant.[26] This was soon used at sea, in combination with lunar distance tables that the British Admiralty published in the *Nautical Almanac* for the year 1768 and after.
Although instrument makers were capable of dividing the graduations of octants and sextants fairly accurately, there was room for improvement in efficiency. In 1770, the English instrument maker, Jesse Ramsden, invented a circular dividing engine. With this, it became possible to engrave very accurate scales on sextants and octants. An instrument which

Portrait of Gerard Pieterz. Hulft, First Councillor and Director-General of the Dutch East India Company. Hulft is surrounded by navigational instruments including a cross-staff. Oil on canvas by Govert Flinck, 1654. Courtesy, Rijksmuseum-Stichting, Amsterdam.

benefitted from this invention was the reflecting circle. This had been suggested by Mayer as early as 1754 but was made practical by the French naval officer, Jean de Borda. It could be used to measure angles up to 180°, and became known as the Borda circle in 1787.[27]

By the end of the eighteenth century, astronomical navigation, with perfected octant and sextant, the chronometer, and lunar distance tables, had reached a level which would not be surpassed until the introduction of radio telegraphy a century later. With these instruments, declination tables, and the Douwes method, a seaman could accurately find his position at sea at almost any time of the day.

Historiography and Sources on the Cross-Staff

In this chapter the published history of the cross-staff to date, and the various other sources essential for the study of the instrument are reviewed. The cross-staff became the subject of historical research during the second half of the nineteenth century. The two most important sources for the study of the instrument are sixteenth to eighteenth-century navigational manuals, and the surviving cross-staffs. Iconography, too, can provide information on the instrument, which is illustrated in a number of examples.

Recent historical studies

The history of the construction and use of the cross-staff became the subject of study and publication in Germany in the second half of the nineteenth century. As W. Köberer pointed out, the reason for this interest was the development of Germany as a colonial power after the 1870-1871 French-German war.[1] Scholarly interest in the history of discovery, geography, and navigation was stimulated by travel to, and contacts with German colonies.

Allegorical representation of the Amsterdam Chamber of the Dutch East India Company. A cross-staff is visible in the foreground. Oil on canvas by Nicolaas Verkolje, after 1702. Courtesy, Rijksmuseum-Stichting, Amsterdam.

Göttingen was an important centre for these studies. This historical activity ended abruptly with World War I, when German imperialist ambitions were smothered. The output, fifteen articles on the cross-staff alone in German journals between 1869 and 1913, was a good basis for further research. Arthur Breusing, who studied in Göttingen and made one voyage as sailor, became a teacher of navigation, and later Director of the nautical college at Bremen.[2] He was the first to publish in 1869 a thorough article on the origin of the cross-staff, touching on Johannes Müller (Regiomontanus), Martin Behaim and the Portuguese contacts with Arabs.[3] This element of the history of the cross-staff, its origin and early use, remained the central theme in later German studies by scholars like Siegmund Günther, Moritz Steinschneider, Albert Schück, Eugen Gelcich, and M. Curtze. The first comprehensive and also very thorough publication on the development of the instrument in later centuries was by the German Master Mariner, Albert Schück.[4] He consulted Portuguese, Spanish, Italian, Dutch, and English contemporary navigational manuals, and also looked at four surviving cross-staffs.[5] Later studies, such as those by Hildemarie Schwindrazheim of the Altonaer Museum in Hamburg-Altona, and M.V. Brewington of the Peabody Museum in Salem, mainly focussed on surviving cross-staffs.[6] They are useful as catalogues, providing material for the Checklist in this book, but give little information on the construction of the instrument. Swindrazheim located a number of staffs in North-German collections and Brewington was inspired by the cross-staffs found in the United States. G.A. Cox, of the Amsterdam Scheepvaartmuseum, wrote a general but thorough article on the history of the instrument, mainly based on the earlier German work.[7]

There are a few general books on the history of navigation that are of importance for the study of the cross-staff. In the late fifties D.W. Waters of the National Maritime Museum, Greenwich, London wrote the so far unsurpassed, standard work, *The Art of Navigation in England in Elisabethan and Early Stuart Times*. It contains a vast amount of information on the use of the cross-staff by English seamen and the contribution of English scholars to its development. In many ways supplementary to Waters' study, and of equal quality, is *Zeewezen en wetenschap* by C.A. Davids. Through this study we now know about the superiority and predominance of the cross-staff over other, contemporary, instruments for the measurement of altitude, mainly in the Netherlands.
The comprehensive study by C.H. Cotter on the history of the navigator's sextant contains a detailed chapter on the cross-staff.[8] The language barrier prevented him from consulting un-translated Dutch sources, resulting in missed information.

Two publications deal with experiments with modern cross-staffs. They give insight into the handling and accuracy of the instrument, information scarcely supplied by contemporary authors. C. St John Hume Daniel, in 1974-1975, sailed with the *Golden Hinde* from Plymouth to San Francisco. During this voyage he carried out a programme of regular observations with a modern mariner's astrolabe, and a cross-staff. The observations were compared with those of a modern sextant.[9] Similar research was done by G. Forty with a replica cross-staff in 1981 and 1982. He concentrated especially on the errors of the instruments and found very valuable results.[10]
To date, only one author, J.J. Roche, has published research on the history of the radius astronomicus, the astronomer's staff, while no study on the surveyor's staff is known.[11]

Contemporary written sources

The sources contemporary to the cross-staff date from between the early fifteenth and the late eighteenth century. They consist of navigational manuals, sea-atlases and archival material. The sea-atlases are well documented by

C. Koeman.[12] A basic list of contemporary manuals, in which the cross-staff is dealt with, was provided in the article by Schück mentioned above. The bibiography of navigational books by H.J.M.W. Peters provides many additonal titles.[13] A check for titles of English manuals[14] revealed nothing not already known through Peters.
The three oldest printed descriptions of the *baculus Jacob*, the astronomer's staff, are by Werner, Apian and Frisius.[15] The first comprehensive description of the cross-staff as an instrument for mariners was by Cortes, in Spanish, dating from 1551.[16] The descriptions by the Spaniard, De Medina, and the Portuguese, Pedro Nunes, although a little earlier, were both very brief.[17] Cortes provided the basis for a number of later descriptions.[18] The first mention in French of the instrument is in the 1554 translation of De Medina's *Arte de Navegar*, but, like the original text, it is very short. The first comprehensive French description of the cross-staff is by Abbé Georges Fournier, in 1643. The earliest known Dutch description of the cross-staff is in the third edition of *Onderwijsinge vander zee / om stuermanschap te leeren* by Cornelis Anthonisz., which appeared in 1558.[19] As well as an illustration, it shows a diagram of the geometrical method to construct the graduation. Bourne, in 1574, wrote the first English description of the instrument in *A Regiment for the Sea*. Although this was based on Cortes' work of 1551 it contained a number of important innovations to the instrument. New developments were also published by Michiel Coignet in 1580, Lucas Jansz. Waghenaer in 1584, and John Davis in 1595.[20] These have been dealt with elsewhere in this book. Waghenaer's *Spieghel* was the first printed sea-atlas, and was translated into Latin, English, French and German. This sort of book, a combination of charts, sailing directions, and a chapter on the art of navigation, was typically Dutch, and dominated European navigation until well into the eighteenth century. Some of the later sea-atlases were by authors like Willem Jansz. Blaeu, Jacob Aertsz. Colom, Hendrik Doncker, Pieter Goos, and the Van Keu-

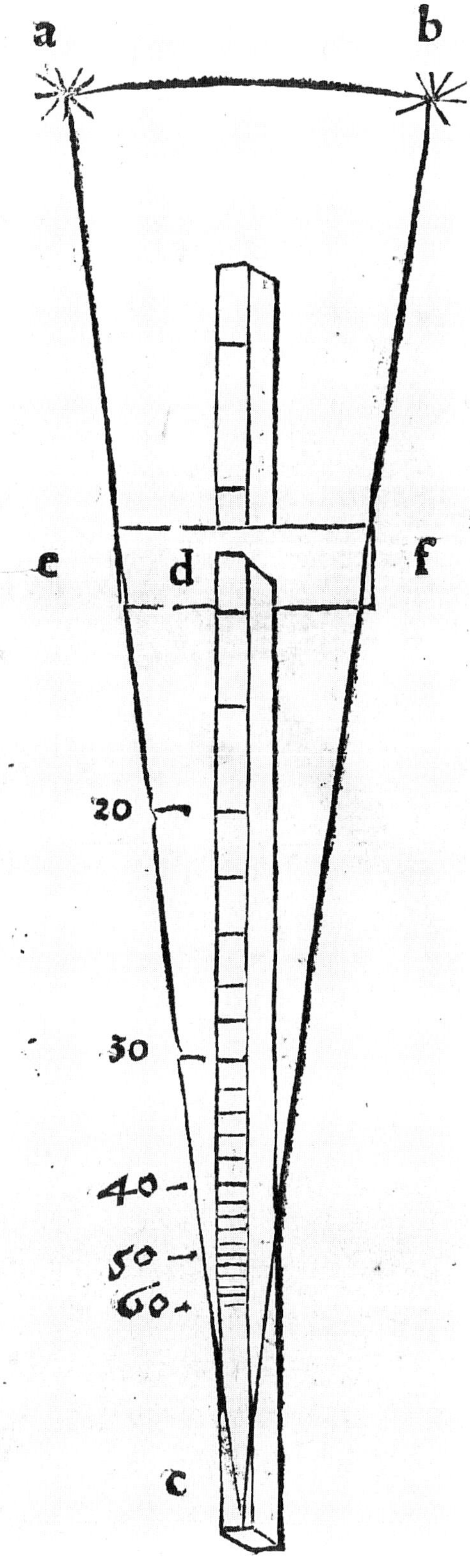

The cross-staff according to Johann Werner, probably the oldest illustration of this instrument.
From I. Verneri, *Nova translatio*, Nurnberg, 1514. Nederlands Scheepvaartmuseum, Amsterdam.

lens. Some, besides appearing in the above mentioned languages, were also published in Spanish and Italian. The chapter or paragraph on navigation always had a description of the cross-staff, the material it was made of, its dimensions, the methods of graduating the scales, and the various ways of using the instrument. The authoritative teacher of navigation, Simon Pietersz., in 1659, did not bother to provide a method of constructing the scales, but simply referred his readers to the works by Waghenaer and Colom.[21]

Archival material on the cartography and navigation of the VOC has been studied and published by Professor G. Schilder of Utrecht University.[22] Much information which he gathered about the sale of cross-staffs to the VOC was made available by him for this book. Of equal importance is the research into estate inventories which include cross-staffs, by Davids, and published in *Zeewezen en wetenschap*.

Iconography

Like the oldest description, the oldest printed illustration of the cross-staff is in Werner's *Nova translatio* of 1514. It was followed, in 1524, by Apian who illustrated how the cross-staff could be used to measure the distance between the moon and a fixed star, for finding the longitude. De Medina, in 1552, provided an illustration of a seaman measuring the altitude of the Pole star. Illustrations such as that given by Rodrigo Zamorano provide information on the graduation.[23] Possibly the earliest coloured illustration of a cross-staff is in the

A Comet over Rotterdam in 1680. Two people are seen observing the comet with cross-staffs. Oil on panel by Lieve Verschuier. Courtesy, Rotterdams Historisch Museum, Rotterdam.

oil painting, 'Allegory of Sight', by Jan Breughel I and Peter Paul Rubens, of 1617, in the Prado in Madrid. The staff, depicted as part of an instrument collection, has only one vane. Allegorical paintings, portraits, and chart cartouches often include a cross-staff, but the representation is not always accurate. By looking at the way the vanes and aperture disk are pictured, one can distinguish whether the instrument used by the artist as a model was an original or not. Sometimes the instrument can be dated, and thus also the illustration. The cross-staff in the allegorical painting of the Amsterdam Chamber of the VOC, for example, must have been after an original, as no painter could know the details of vanes and aperture disk shown. But the shape of the vanes on the 1676 vanitas painting by Adriaen Valck has never been found on surviving staffs. This shape is, however, often found on the usually inaccurate and superficially drawn cartouches of charts.
The number of vanes and the presence of one or more aperture disks and a bone vane, can be an indication for the dating of the illustration.
Sir Jonas Moore, Surveyor General of H.M. Ordnance and Fellow of the Royal Society, in 1681 recorded the use of the cross-staff to measure the distance between two stars. This may not have been as theoretical as its seems. A painting by Lieuve Verschuier, of 1680, shows a comet over Rotterdam being observed using cross-staffs.

A vanitas painting including a cross-staff. Oil on canvas, by Adriaen Valck, 1676. Courtesy, Hamburger Kunsthalle. Photograph Elke Walford.

The cross-staff as source

Historians, unlike writers of contemporary literature, hardly used the instrument itself as a source. No comprehensive study at all was made of the stamps, the maker's names and initials, numbers, or the dimensions of staff and vanes, and the aperture disk. The little that was published on the scales was based on contemporary literature, and not on actual instruments. This gap in the study of the cross-staff can be explained.
In order to draw useful conclusions from this instrument, or any instrument for that matter, and to compare these with contemporary printed sources, a large number of instruments should be studied. No author to date has seen nearly as many as are in the Checklist in this book. Schück mentioned only four cross-staffs, Schwindrazheim and Brewington each registered ten, and Cox three. A source which recently became available, and was not at the disposal of earlier scholars, is the result of maritime archaeology. Ship-wrecks have yielded several cross-staffs providing information that was hitherto unknown.[24]

The data on the 95 cross-staffs in the Checklist is the last source to be mentioned. It has been compared with the data found in contemporary written sources in the last chapter of the introductory part of this book.

Introducing the Cross-Staff

Name and origin

The subject of this book, the mariner's *cross-staff*, was originally known as *baculus Jacob*. It was also known by the English as *ballastella*, *Jacob's staff* or *fore-staff*. The latter name was introduced by Henry Gellibrand, possibly around 1634, to distinguish it from the back-staff.[1] The French sometimes called it *baton de Jacob* and the Italians *scala di Jacob*. The Spanish name *balestilla*, the Portuguese *balhestilha*, the more commonly-used French *arbaleste* and *arbalestrille*, and the Italian *balestriglia*, were given to the instrument as, when in use, it resembled a cross-bow.[2] Cox suggested that the seaman's expression 'to shoot the sun' originates from this name.[3] Besides *baculus Jacob*, the Dutch, in the sixteenth century, named it *staf baculus* but, in later centuries, more generally *graadboog*, *graadstok* or simply *stok* and sometimes, *Jacobsstaf*. The German *Gradstock*, and the Swedish *grad-bogar* or *grad-ståcker* and the Russian *grádbog*, were probably translations from the Dutch.[4]

The English mathematician, Thomas Hood wrote in 1590[5] that some believed the name *baculus Jacob* was based on Genesis 32:10. This says 'I [Jacob] am not worthy of the least of all the mercies, and of all the truth, which thou hast shewed unto thy servant; for with my staff I passed over this Jordan; and now I am become two bands'. It has also been suggested that it was so named as it resembled the biblical Jacob's ladder, between heaven and earth.[6] A third suggestion based on the Old Testament was that an early, coarsely engraved staff resembled partially stripped rods of wood.[7]

It is generally accepted by historians that the baculus Jacob was first described around 1342 by the Catalan-born Jewish philosopher and scientist Levi ben Gerson (1288-1344) who, at the time, was living in Avignon. He may even have invented the instrument, which was then only used by astronomers. Regiomontanus and his patron, Bernhard Walther knew of Levi ben Gerson's treatise and Walther used a *baculus Jacob* in Nurnberg between 1476 and 1504.[8] The step towards making it available for mariners was, perhaps, set by the astronomer Martin Behaim (Nurnberg *c.* 1459-1507 Lisbon). He was a pupil of Regiomontanus, who went to Lisbon in 1484 and contributed to Portuguese voyages along the West-African coast in the years 1484-1485. Günther suggested that Behaim introduced the *baculus Jacob* to Portuguese explorers and thus made it available for use at sea.[9] There is, however, no record of it being used by the Portuguese at sea in the fifteenth century and the Behaim theory has yet to be proved.[10] It has been more broadly accepted that the adaption of the *baculus Jacob* for use at sea was inspired by the *kamal*. This was an instrument for the measurement of altitude at sea, used by Arabs and seen by Vasco da Gama in 1498. The *kamal* consisted of a rectangular tablet of wood at the centre of which a cord was secured. To measure the altitude, the tablet was held with its upper and lower edges in coincidence respectively with the celestial body, and the horizon vertically beneath it. The cord was held taut either at eye-level or by the teeth. The length of the cord between tablet and eye or teeth was then a function of altitude of the body.[11]

The earliest reference to the mariner's cross-staff in a navigational work dates from shortly before 1520.[12] The earliest comprehensive printed description is by Marten Cortes in *Breve Compendio de la Sphera y de la arte de Navegar*, from 1551, while the oldest published illustration is to be found in *Regimento de Navegacion*, by Pedro de Medina, which appeared in 1552. Both authors write that it was used for observing the altitude of the Pole star or any other star. For the sun's altitude the cross-staff was of no use, unless the sky was covered with a thin layer of cloud (to prevent the observer from being blinded by sunlight, see below), and the horizon was absolutely clear.

Other staffs

There were a number of staffs related to the mariner's version, which should be briefly introduced. In the first place the mariner's cross-staff should be distinguished from the astronomer's staff or *radius astronomicus* and the surveyor's staff or *radius geometricus*. Older than the mariner's version, these two were no longer in use by the end of the seventeenth century.[13] Also, the latter two are more rare. There are two known astronomer's cross-staffs, in Madrid and London,[14] and two possible surveyor's staffs, now preserved in Florence.[15]
The cross-staff, by which hereafter the sea or mariner's version is meant, was often used in conjunction with other instruments for altitude measurement. The older nautical quadrant and mariner's astrolabe have already been mentioned in a previous chapter. In later years a number of instruments were derived from the cross-staff. They all had in common that they were used with the observer's back towards the sun, to eliminate having to look into bright sunlight. The principle was described by the English astronomer Thomas Hood in 1590, but the instrument he designed for that purpose was not very practical in use.[16] The most famous was the back-staff or Davis quadrant named after the English seafarer, John Davis, who first described it in 1595 in *The Seamans Secrets*. In the early form it consisted of a staff with a sliding transom. With his back towards the sun, the observer had to slide the transom until its shadow coincided with a vane at the end of the graduated staff. The transom on the later, more sophisticated, back-staff became a graduated arc, fitted with sliding vanes. The English quadrant, as the back-staff was sometimes named by the French and Dutch,[17] was popular on both sides of the Atlantic Ocean, and remained in use until the end of the eighteenth century. It is also the only type of which a large number of examples are known to have survived.[18] Another variation which was used at sea, although for a shorter period than the back-staff, was the *spieghel-boog* [mirror-staff]. It was patented in 1660 by the Dutchman, Joost van Breen, and described by him in *Stiermans Gemack*, published in 1662. It resembled a cross-staff but had a mirror placed on a short vane in which the image of the sun and the end of a long vane were reflected. If this image coincided with the directly observed horizon, the altitude of the sun could be read from the scale on the staff. The instrument was provided to the ships of the Zeeland Chamber of the

A list of various cross-staffs, a coloured glass, and a 'mirror-staff', with their retail prices. Detail from a sale catalogue for instruments, made by the firm of Van Keulen, dated 1777. From an appendix in K. de Vries, *Schat-kamer*, Amsterdam, 1777, in the Library of the Nederlands Scheepvaartmuseum. Nederlands Scheepvaartmuseum, Amsterdam.

GRAAD-BOGEN in zoorten, als dito met dubbelde Bosſen, dubbelde Vizieren en dubbelde Teekening. - - - f5:-:-

Dito, met dubbelde bosſen, Vizieren en Teekening. - - - - f3 10:-

Dito, met ordinaire Bosſen, Vizieren en Teekening. - - - - f2-10:-

Dito, zonder Bosſen, met Vizieren. f2:-:-
en tot minder Pryzen, zo die aanbeſteed worden.

Gezicht Glaasjes, tot dito. - 6 ſtuiv.

Spiegel Graad-Bogen, - - - f4-10:-

PEIL-

VOC, after 1670. Although no examples of the mirror-staff are known to have survived, it could still be purchased in Amsterdam more than a century after its invention.[19]
Far less is known of instruments named *crommeboogh* [curved staff], *knipboog* [snap-staff] and *cruysboog* [cross-bow]. Of the *crommeboogh* we do know that it was invented in *c.* 1617 by Maerten Joosten Wagenaer, a boxmaker in Enkhuizen.[20] Of the *cruysboog*, mentioned in 1617, we know that it cannot have been the instrument of that name described by Edmund Gunter in 1624.[21] Robert Dudley, in 1661, mentioned a variant of the cross-staff, the invention of which he attributed to a seaman named Abraham Kendall.[22] The Frenchman, C-F Millet Deschales, described in 1677 a cross-staff with apertures and a sight-vane.[23] Finally, in the category of staffs of which no example is known to survive, Joseph Harris in 1730 published a proposal for a 'new Fore-staff', supposedly twice as accurate as the 'common' fore-staff.[24]

Use

The basic elements of a cross-staff, a four sided staff and one or more vanes, are made of wood. The staff is square in cross-section, and has four smooth sides. The average length is about 800 mm and the sides, on which graduation scales are cut, are about 16 mm wide. The vane, shorter than the staff, is flat and about 40 mm wide. In the middle it has a square hole just large enough for the staff to pass through, so that the observer can slide the vane along the staff.
In order to 'shoot', the eye-end of the staff was placed near the observer's eye and the other end was directed at a point in the sky approximately half way between the horizon and the celestial body. The vertically placed vane was then slid along the staff until its upper edge 'touched' the celestial body. At the same instant the vane's lower edge had to 'touch' the horizon. The vane was then clamped to the staff and the altitude read from the scale.
The early sixteenth-century descriptions of the cross-staff are in Spanish and Portuguese, when it was mainly used to measure the altitude of the Pole star. By 1574, English seamen at least also used the cross-staff to measure the altitude of the sun. In that year the influencial English teacher in the art of navigation, William Bourne, published *A Regiment for the Sea*. He described how the cross-staff could be used for the altitude of the sun, observing either the upper or lower limb. Another novelty described was a piece of a coloured glass which, placed at the end of the vane, would prevent the observer from being blinded by sunlight. In the event that no coloured glass was available, Bourne advised to observe the sun's upper limb, covering it with the vane's end, and subtracting the semi-diameter from the measured altitude. This, then estimated at 15 minutes of arc, was until *c.* 1595, the only correction applied to cross-staff observations. There is no contemporary evidence to indicate which of the two limbs of the sun was best to observe, but observations with a modern cross-staff gave better results using the upper limb than the lower.[25] Bourne advised not to use the cross-staff if the sun's altitude was more than about 50°. The larger the angle, the larger the intervals between the degrees on the staff, and, with larger angles the blinking distance would increase the error.
The 'blinking' of the eye between the upper and lower ends of the vane was necessary to ascertain if it was in the correct position. It could be avoided when, in 1595, it became possible to use the staff backwards, that is with the observer's back to the sun, as suggested by Davis. There were two ways to observe thus, the normal way and the Dutch fashion, so named as it was popular with Dutch seamen.[26] In the normal way the shortest or horizon vane was fixed at the end of the staff and the longest vane could slide. The observer, with his back to the sun, would look along the lower edge of the long vane, aligning that with the middle of the horizon vane and the apparent horizon. The long vane would then be slid along the staff until the shadow of

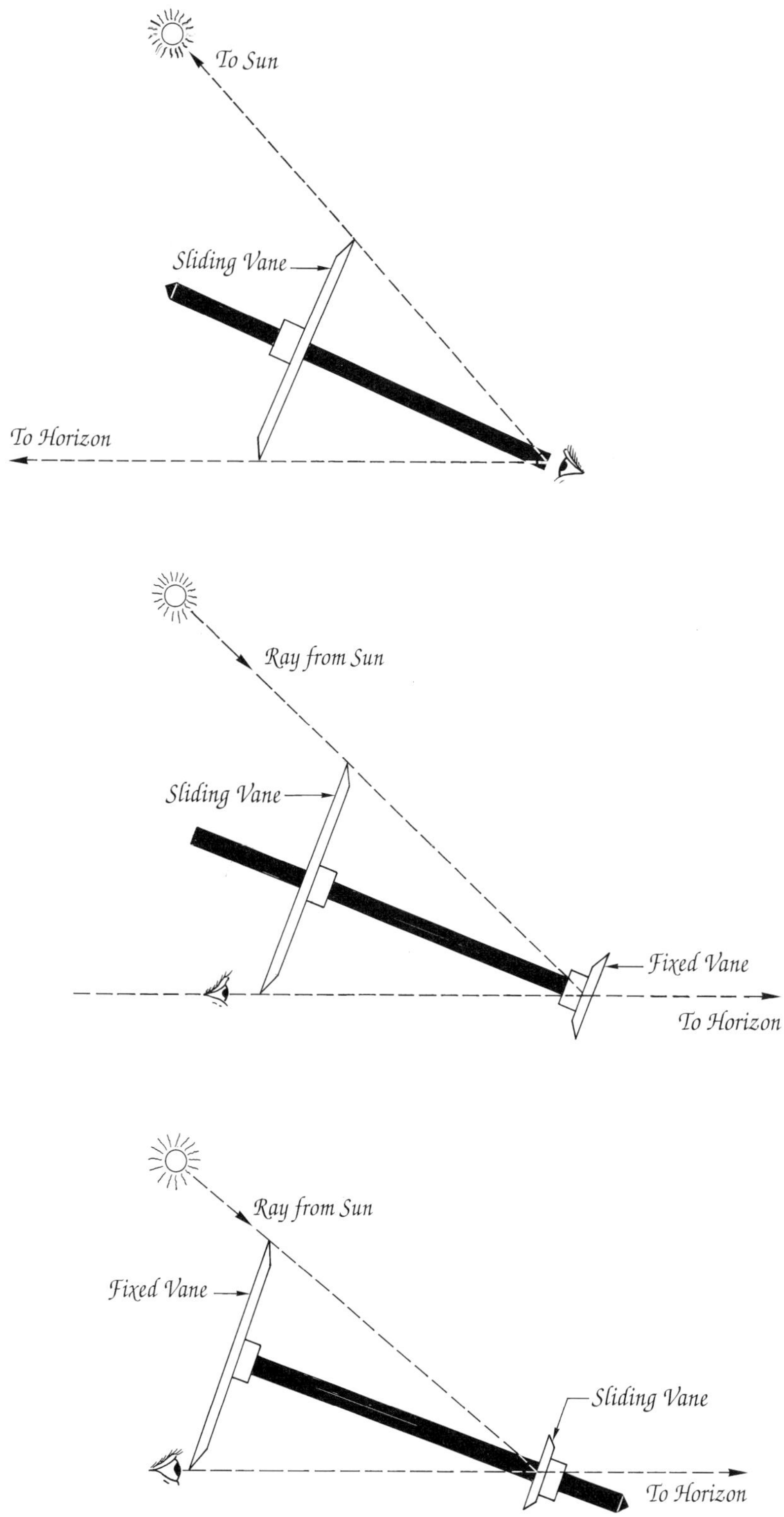

The three methods of using the cross-staff. A is the forward method, B the normal backward method, and C the backward method according to the Dutch fashion.

its upper edge fell on the middle of the horizon vane. At that instant, the sliding vane was held tight to the staff, and the angle read from the scale, corresponding to the long vane. In the Dutch fashion, the longest vane was fixed at the extreme eye-end of the staff, and the horizon vane could slide. The observation went in the same way, but now the angle was read on the scale for the short vane. Contemporary authors describing both methods did not distinguish between their quality. However, observations with a modern cross-staff, on the *Golden Hinde* in 1974 and 1975, have shown that the normal method was clumsy compared to the Dutch fashion. It is easier to control the sliding vane when it is short, while, at the same time the long vane, when pressed against the body, is less influenced by the wind.[27] It was also found, that the backward method is more accurate than observing while facing the sun.[28]

One theoretical aspect of the backward method that should be mentioned here, was described by Sir Jonas Moore, in 1681. He appears to be the first and only one to suggest that, in the event of a thin layer of clouds, a lens or 'small convex glass' should be fitted at the end of the vane. This, when observing backwards, would converge sunlight and cast a small bright spot on the horizon vane. The idea was, no doubt, inspired by John Flamsteed who, around 1670, suggested the use of such a lens on backstaffs.[29] The 'Flamsteed lens' is frequently found on these instruments; for cross-staffs, however, there seems to have been no following.

Errors

By taking a backward observation, looking into bright sunlight was avoided and also the error caused by ocular parallax was eliminated. This was the main instrumental error of the cross-staff and was first described in 1574 by Bourne. If a cross-staff was correctly made, the lines of sight along the ends of the vane coincided in the middle of the flat eye-end of

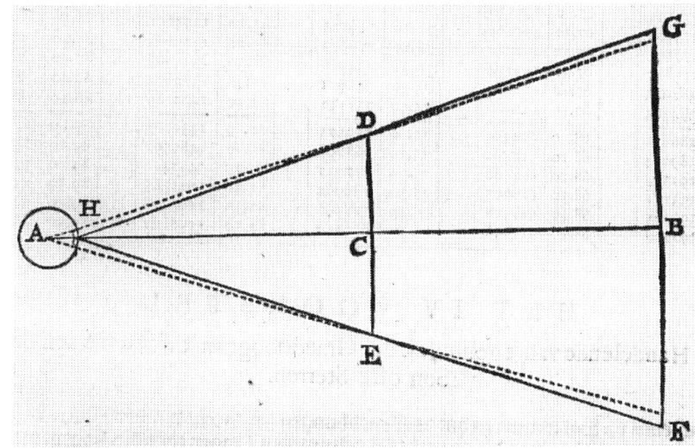

Illustration showing ocular parallax.
From A. de Graaf, *De Seven Boecken*, Amsterdam, 1658. Nederlands Scheepvaartmuseum, Amsterdam.

the staff. However, to obtain a correct observation, they should fall on the eye's retina, slightly beyond the staff's end. In order to accomplish this, the staff should be placed as near to the eye as possible. But, success depended very much on the observer's facial anatomy and how the staff was positioned on it. Some authors advised to place the staff below the eye,[30] others wrote that it should be beside the observer's eye, either on the inside or outside.[31] Some wrote that the exact placing did not matter.[32] Modern research has shown that, nearly always, the staff would be placed too far forward, resulting in a higher reading than the altitude actually was.[33]
To compensate for the error created by ocular parallax, Bourne advised to 'pare away a little ende of the staffe', in other words to shorten the staff by sawing off a small part at the eye-end. In that way, it was hoped that the sight lines would coincide 'beyond' the end of the staff and fall on the retina.[34] This 'correction', which amounted to a degree or more, was strongly opposed by Dutch scholars after Bourne, such as Blaeu.[35] They pointed out that the result was more often an increased than a reduced error. Georges Fournier, who in 1643 wrote an authoritative book on navigation, claimed the shortening could amount to as much as $2^1/_2°$, and was done by both Dutch and French seamen. According to him, some makers, to save the seamen the trouble of shortening the staff, allowed for the $2^1/_2°$ correction in the graduation.[36]
Around 1595, a method for the observer to measure his ocular parallax became available.

It was described by Davis and involved the use of two vanes and consequently two scales.[37] The vanes should be slid on to the staff, parallel to each other, and fixed for the same angle on their respective scales. With both ends of the two vanes in line, the observer should memorize the position of the staff's eye-end in relation to his eye. That position, which differed for each observer depending on his facial anatomy, should always be assumed when taking an observation. This method of finding one's ocular parallax was published in almost every later navigational manual, with an illustration. Observations with a modern copy of a cross-staff have shown that this method, although theoretically correct, is in practice extremely difficult to apply.[38]
Fournier, in 1643, provided three different methods of checking the cross-staff and finding the error. The first was by comparing a measured angle simultaeously with that on an astrolabe. The second was by measuring with all three vanes in quick succession, and comparing the three readings. Both these methods, if performed by a single observer, could only be a success if the measured altitude remained near to constant for the duration of the experiment. The third method could always be applied but required a long and straight ruler. As with Davis' method, the vanes were fixed to the staff for the same respective angle. Instead of looking along the ends of the vanes, the cross-staff was laid flat on the deck, and the ruler placed against the ends of the vanes. If it also touched the staff in the middle of the eye-end, it was without error.
The English author, Joseph Harris, in 1730, wrote that ocular parallax could be eliminated by observing the celestial body and apparent horizon along the upper and lower end of two vanes, ignoring the position of the staff's eye-end. The altitude could be read on either of the scales. Harris admitted that this was a difficult observation.[39]

Besides ocular parallax, there were random errors which were unavoidable. These were the influence of wind, movement of the ship, the graduation of the staff, the declination tables, and refraction because of using non-plain-parallel coloured glass to protect the eye against sunlight.[40] The latter error was known as early as 1584.[41] Other corrections, such as refraction, dip and eye-height, should also be taken into account of any measured altitude. They became available in tables which appeared in the 1590s but will not be dealt with here extensively, as they are not specific for the cross-staff. On the basis of observations with a modern cross-staff, it has been estimated that the total error in a cross-staff altitude measurement was about 20 minutes of arc.[42]
In the middle of the seventeenth century, expedients were developed to improve the accuracy of the backward observation. They were a brass aperture disk, named the 'Dutch shoe', attached to the end of a vane; and a wood or bone vane, simply called the 'bone', attached to the horizon vane. They will be more extensively dealt with in the chapter on the comparison. Towards the end of the seventeenth century, the vanes were also fitted with clamping screws, so that, by 1700, the cross-staff was fully developed and had become the most accurate and reliable instrument available for measuring altitudes at sea.

Graduation

Until the end of the sixteenth century only one side of the staff was graduated and, consequently, only one vane was required. The graduation, from 0° to 90°, had to be compressed on to one scale. This meant either a long staff or small degrees. The long staff was typical for astronomers, while at sea, for practical reasons, a shorter staff was preferred. For use at sea, around 1580, it became the practice to make cross-staffs with three scales, and consequently three vanes, each representing a section of the 0° to 90° graduation. This improvement of the cross-staff was first described in 1580 by the Flemish teacher in the art of navigation, Michiel Coignet. His *Nieuwe Onderwijsinghe* was an appendix to the Dutch translation of De Medina's *De Zee-*

vaert, by then almost thirty years old. Coignet wrote that scales on three sides of the staff were already being used by all capable pilots.[43] He also numbered the sides of the staff. Side one, with the longest vane, was for angles over 30°, side two was for angles between 10° and 30°, and side three, with the shortest vane, for angles under 10°. With three supplementary scales greater accuracy of observation was possible, as the degrees on the different scales were larger than on one scale.[44] Four years later, in addition to the altitude scale, the Dutch pilot, Waghenaer, in the *Spieghel*, introduced a scale for zenith distance. His cross-staff had two graduated sides, one with a scale for zenith distance, to measure the complement of the sun's altitude, the other to observe the altitude of the Pole star (observer's latitude). With a scale for zenith distance, the calculation of latitude from the sun's altitude had been simplified.[45]
Another Dutchman, Jacob Aertsz. Colom, author of *The Fierrie Sea-Columne*, wrote in 1633 that staffs were being made with scales for both altitude and zenith distance on each of the three graduated sides.[46] In the 1650s all four sides of the staffs had graduations, which was first described by the Amsterdam teacher in the art of navigation, Abraham de Graaf.[47]

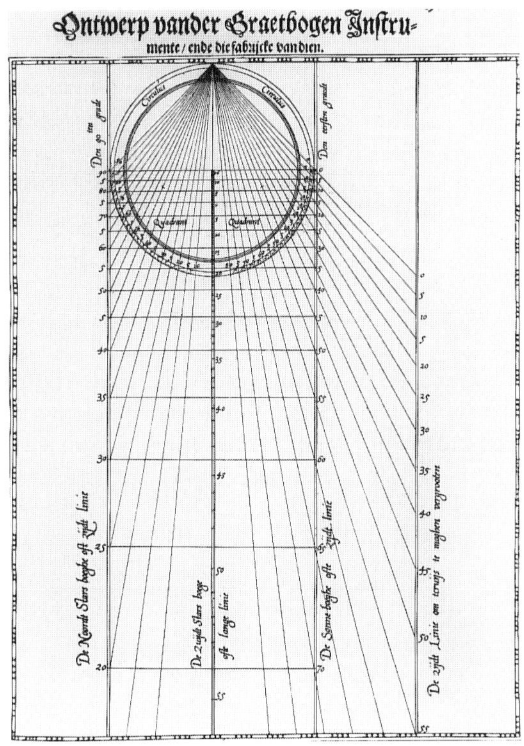

The geometrical method of constructing the graduation, by Lucas Jansz. Waghenaer of Enkhuizen. From L.J. Waghenaer, *Spieghel der Zeevaerdt*, Leiden, 1584. Nederlands Scheepvaartmuseum, Amsterdam.

There were several methods for applying the graduated scales to a staff. Werner, in *Geographia*, of 1514, was the first to describe the geometrical and the mathematical method; for the latter he referred to the tangent tables of Regiomontanus.[48] Moore also mentioned a diagonal method, which required a table of natural cotangents.[49] The mathematical method was regarded as the more accurate of the two. Dirk Makreel who, in 1672, was appointed examiner of the lieutenants of the Amsterdam Admiralty,[50] opposed the geometrical method as it could create errors.[51] Nevertherless, both methods for graduation remained available in manuals until well into the eighteenth century.[52]
A third way to graduate a staff, less advertised than the previous two, was to copy it from another staff. Blaeu, and after him Adriaan Metius, warned against this custom without having checked the original for errors.[53] The copying was condemned by Makreel.[54] It was described by Claes Jansz. Vooght, a mathematician, surveyor, and teacher in the art of navigation, in 1684.[55] The two staffs, original and blank, should be placed rigidly together so that, using a square and a knife, the scales could be copied.
In order to graduate the scales according to the geometrical method, a construction diagram had to be made.[56] This was a flat piece of board, covered with pasted-on paper, and rectangular in shape. The long side, the base line, had the length of the intended staff, and the short side that of half the length of the longest vane. The lower left corner of the board represented the eye-end of the staff, and was the centre of concentric circles of which the radius equalled half the lengths of the intended vanes. That is why the eye-end of the staff was also named centre. From this centre, using compasses, an arch of 45° was set on the base line.

Next, this arch had to be divided into 90 equal parts, in other words in angles of $^{1}/_{2}$°. This was done by first dividing the arch in three angles of 15° and then dividing each of those into three angles of 5°. These 5° angles were then divided by two, leaving angles of $2^{1}/_{2}$°, and finally these were divided into five angles of $^{1}/_{2}$° each. For every $^{1}/_{2}$°, straight lines were drawn from the centre. Then, from the centre upwards, the half lengths of the vanes were marked on the left side of the diagram. From these marks, lines were drawn across the board from left to right, and parallel to the base line. The blank staff was then placed on these parallels, and where the $^{1}/_{2}$° lines crossed the parallel, the staff would be marked with a knife. Divisions in less than 30 minutes were probably found with the compasses.

Although the length of the vanes could be measured from the diagram, some authors advised to make the vanes first and to project their half lengths on to the diagram at the very start, eliminating the risk of a difference in length at the end of the process.[57]

The mathematical method was laborious but more accurate.[58] Again a pasted board was required, on which the long base line represented the intended staff. Using compasses, the half vane lengths were projected on the base line, from the centre at the left, towards the right. With compasses, each vane length on the board was divided into a number of equal parts.[59] In a tangent table, provided especially for this purpose in most navigational manuals, the number of parts for each degree could be found and marked on the diagram, and from that pricked on the respective sides of the staff, with the compasses. The most difficult element of this method, and an obvious source of errors, was the dividing in equal parts. It had to be done three or four times, depending on the number of vanes required. However, once a maker had constructed a diagram, he could use it to make more staffs.

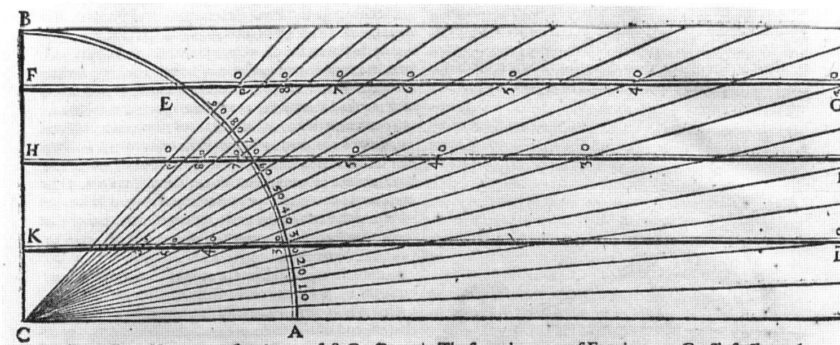

The 23 *Point.* Hou you muſt make a perfect Croſſe-ſtaffe, and how you ſhall marke it.

The Croſſe ſtaves are oftentimes made the one after the other by imitation, or patterns, and that many times without judgement or knowledge whether the pattern bee good or not, whereof notwithſtanding a man ought to bee very certaine, for it is a matter of great importance and therfore I have here ſet downe two ſeverall wayes eaſie to be underſtood, how to make the ſame exquiſitely and perfectly upon good reaſon & good ground.

Make an even ſmooth board, and paſte good paper upon it, then draw a ſtrait line upon one edge or ſide thereof, as you ſee in this figure a line marked C A D, then marke another line as you ſee it marked B C, that B C A bee a right and juſt halfe ſquare, then take a good paire of Compaſſes, and ſet one foot thereof in the angle marked C, and compas there with as you ſee B E A, ſo that A E B may bee a right quadrant, or fourth part of a Circle, then devide it into two parts, to E, and then devide the other halfe (which is A E) into 90 parts, or degrees as thus: firſt devide it into 3 parts, and then devide each 3 parts again into 3, and then it wil be divided into 9 parts, which done, devide each of them into two parts, and thoſe diviſions or parts divide into 5 parts, and then it will bee divided into 90 parts: then take a ſtraight line: and lay the one end thereof upon the Center C, and ſo forth upon every ſeverall point, which you have divided in the Quadrant, and then draw lines from the Center C through all the foreſaid points, as long as the board will containe them, as you may ſee in this figure: which done then your inſtrument is fully made and prepared to marke your ſtaves there upon. Which to doe, with your compaſſes take juſt the half length of your Croſſe, and ſet the one foot thereof in the Center, and put the other at F, and doe the like from D right to G, which two points or pricks draw with a ſtraight line unto each other as you ſee in the line F and G, and then mark where the line F G is cut through by the lines that come out of the Center through the degrees or points of the quadrant: for thoſe cuttings through, ſhew upon the ſaid line the right marking of your Croſſe-ſtaffe: that is, you muſt marke your Staffe juſt as you ſee that theſe lines in F G are marked, by the lines that come out of the center and paſſe through it, whether the Croſſe bee long or ſhort, you muſt follow the ſame order that you are taught in the figure, the line H I is drawne for a ſhorter Croſſe then the former, and K L for one that is ſhorter then that, viz whoſe halfe is juſt ſo long as the Line C K.

The ſecond manner of Framing your Croſſe-ſtaffe much excelleth the firſt in perfection and curioſity, ſo you divide it perfectly with your Compaſſes, which is done in this manner. Upon an hard even board which is paſted with paper draw a ſtraight line as long as your ſtaffe is, and with point compaſſe take the juſt halfe or the length of your Croſſe, whereunto you mean to mark your ſtaffe, and prick it many times along in the ſaid line as you can: & divide each of the ſaid lengths into 1 1000 equall parts, then look upon the table here after inſuing, which ſhewed you how many of thoſe points or parts you ſhall marke for each degree, and that you muſt doe in manner following, from the end of the ſtaffe, which you will make the eye end pricke juſt the halfe length of the Croſſe, and there make a Croſſe ſtroke, from whence prick for every degree ſo many of the aforeſaid parts as the table enſuing ſheweth you. As for example, for the firſt degree prick 176 parts, for the ſecond 355 parts, for the third 538 parts, for the fourth 724 parts, and ſo forth: Remembring that you muſt prick all the lengths of pricks or parts aforeſaid, from the aforeſaid croſſe ſtroke, which is marked upon the halfe length of the Croſſe into 1 1000 parts, then devide the whole Croſſe ſo, and then take for every degree halfe ſo many as the Table ſheweth unto you. And if you cannot doe ſo, devide halfe the Croſſe into a 1000 parts, and then leave the hindermoſt letter of the table out, as where you ſhould for the firſt degree take an 176 parts, take but 17 parts, leaving the laſt figure out: but if your croſſes be ſo ſmall that you can devide the halfe of them into no ſmaller then an hundred parts, then leave out the two laſt figures in the table. But you muſt underſtand that the more parts that you devide the croſſes into, your markes will fall out ſo much the better & perfecter.

de.	parts.	de.	parts.	de.	parts.	de.	parts.	de.	parts.
1	176	19	4019	37	10057	55	21716	73	56612
2	355	20	4181	38	10503	56	22708	74	61154
3	538	21	4550	39	10965	57	23759	75	65958
4	724	22	4826	40	11445	58	24874	76	71445
5	613	23	5108	41	11943	59	26059	77	77769
6	1106	24	5399	42	12460	60	27321	78	85144
7	1303	25	5697	43	12998	61	28667	79	93854
8	1504	26	6003	44	13558	62	30108	80	104301
9	1708	27	6318	45	14142	63	31663	81	117062
10	1918	28	6643	46	14751	64	33315	82	133007
11	2131	29	6976	47	15386	65	35107	83	153469
12	2349	30	7320	48	16051	66	37046	84	110811
13	2572	31	7675	49	16746	67	39152	85	219038
14	2799	32	8040	50	17475	68	41445	86	276362
15	3032	33	8418	51	18239	69	43955	87	371885
16	3270	34	8807	52	19042	70	46713	88	561810
17	3514	35	9210	53	19887	71	49758	89	1135891
18	3764	36	9626	54	20777	72	53137	90	infinit.

G

The

The description of the geometrical and the mathematical methods of graduating a staff, by Pieter Goos. From P. Goos, *Lighting Columne*, Amsterdam, 1662. Nederlands Scheepvaartmuseum, Amsterdam.

Production and Diffusion of the Cross-Staff

In this chapter the production of the cross-staff will first be analysed, after which the diffusion of the instrument will be discussed.

Production

Early contemporary descriptions of the cross-staff in navigational manuals for seamen do not refer to the makers of the instrument. The fact that most descriptions include an elaborate instruction on how to make a cross-staff, in particular the scales, suggests that they were generally made by the seamen themselves. On the other hand, the advice by Bourne in 1574 to his readers, who were seamen, to shorten the staff in order to eliminate the ocular parallax, could be an indication that they were made by craftsmen, i.e. professional instrument makers.[1] The earliest known name of a professional cross-staff maker, Francis Cooke, is found on the title page of Thomas Hood's 1590 description of the instrument.[2] The pilot and cartographer, Albert Haeyen, in 1600, wrote how he, together with surveyor, Adriaen Ockersen, improved the construction of the cross-staff's graduation. He added that one should not have them made by inexperienced makers and that in England the construction was well seen to.[3] The oldest reference to the availability of a standard cross-staff, as an example, dates from 1606 in the Casa de la Contratacion.[4]

It is not unlikely that the first professional cross-staff makers were craftsmen, such as carpenters. In 1620 a boxmaker, by the name of Pieter Willemsz., provided cross-staffs for the VOC.[5] An early mention of professional cross-staff makers was by Georges Fournier and dates from 1643. He wrote that these, in the French port of Dieppe, used tangent tables to construct the scales according to the mathematical method. He also referred to a discussion between the Dutch cartographer, Colom, and cross-staff makers.[6] The former, according to Fournier, accused seamen of wrongly applying corrections such as eye-height and ocular parallax. Fournier, however, blamed the makers of shortening cross-staffs in order to eliminate ocular parallax. He claimed that, out of every 500 cross-staffs in use in France and the Netherlands, not a dozen were without error. Fournier was convinced that this was the cause of the many faults that were made in position-finding, and proposed to solve it by having the government commission 'standard' cross-staffs. Those already in use at sea should be compared with these standards, and, if in error, be confiscated. It has not been possible to establish whether the discussion between Colom and the makers was published. It was not found in his sea-atlases.[7]

Hendrik Doncker, in 1655, was the first to advertise as a cross-staff maker, on the title page of his *De Lichtende Columne*. Several compilers and publishers of sea-atlases, at least in the Dutch Republic, were also cross-staff makers. From the time that Doncker was active, numerous cross-staff makers are known to have worked in Amsterdam.[8] The fact that, in this city, so many makers' names are known, and hardly any in other seafaring countries, suggests that Amsterdam was an important centre of the production. This is supported by a number of facts. The Swedish Admiral, Werner von Rosenfeldt, in 1693, wrote that diagrams for constructing the scales of a cross-staff, could be purchased, particularly in Holland.[9] There is another indication of Dutch predominance in cross-staff making. As we have seen, English seamen named one of the two methods of using the cross-staff with the back towards the sun, the Dutch fashion. It had been developed and used by the Dutch and subsequently recommended by the English author, John Seller.[10] The same happened with the 'Dutch shoe' a brass aperture disk, so named because it was first used by the Dutch. It is dealt with in the chapter on the surviving cross-staffs.

By the time this edition of Seller's book was published, Johannes van Keulen, book seller, chart and instrument maker, was active in Amsterdam. Not only are the majority of the surviving cross-staffs made by him and his family, but we know that, between 1731 and 1748 they sold 1148 cross-staffs to the VOC alone.[11] These figures do not include sales to other branches of shipping, such as the trade with the Americas, the Baltic, and whaling in the Arctic. In fact, it seems apparent that the Van Keulens were unable to meet the whole of the demand for cross-staffs in the eighteenth century. During the same period, other makers, such as Johannes Loots, Jochem Hasebroek, and Lambertus Dankbaar, were also independently active in Amsterdam. Finally, the large number of surviving Dutch staffs and the very few from other countries, suggest the predominance of the Dutch Republic in this trade. Around 1750, authors of navigational manuals ceased to publish descriptions of the cross-staff, and even denounced its use in favour of the octant. Nevertheless, the production of cross-staffs continued, and for a considerable period too. Even the Van Keulens, who were producing octants as early as 1744 and selling them to the VOC in 1751,[12] kept on advertising cross-staffs and, as is evident from the surviving examples, producing them until the end of the century.

Diffusion

As we have seen in previous chapters, astronomical navigation was first introduced in Spain and Portugal in the late fifteenth century. The introduction of the cross-staff followed around 1515. As the knowledge of the art of navigation spread North, further into Europe, so did the use of the cross-staff. Outside Spain, English and French seamen were probably the first to use the instrument. This is supported by two early references. The first concerns an action in the High Court of Admiralty in 1533 about the loss of John a Borough's sea chests. It revealed that Borough had, among other instruments, a cross-staff.[13] The second reference, dating from 1540, is a report of the taking by the *Barbara* of London of a Spanish ship with on board a French pilot from Dieppe. The English found, among other instruments belonging to the pilot, '... a balestely, an instrument belongyng to the office of a pilotte for the night ...',[14] indicating its specific use for the altitude of the Pole star. The Dutch were using it around 1558 and, by the end of the century, were improving the instrument. As Davids has shown, after 1550, Dutch nautical knowledge spread to north-eastern Europe through navigational manuals and sea-atlases.[15] There can, therefore, be little doubt that these sea-atlases, which were translated and used all over Europe, contributed to the diffusion of cross-staff. This is supported by the fact that several of the surviving Dutch cross-staffs, now in German and Scandinavian collections, originate from local seafaring families, or were found in local ship-wrecks.

In various manuals the cross-staff is described as popular and much used. The English scholar, Thomas Blundeville, in 1594, wrote that it was 'commonly used in these dayes',[16] Edmund Gunter, in 1636 called it 'well known to seamen'[17] and Andrew Wakely in 1664 'a common and ordinary instrument for seamen'.[18] According to Moore, in 1681, English seamen preferred the cross-staff above the various other instruments for the measurement of an altitude. Von Rosenfeldt, in 1693, described it as the most important of instruments for altitudes at sea.[19] Seller, in 1705, wrote of the cross-staff that, although it was an instrument 'of some antiquity in navigation', it performed with sufficient exactness. There can be no

Rear-Admiral Willem van der Zaan (1621-1699). On the left a cross-staff, set for the backward observation according to the Dutch fashion, is visible. The long vane is fixed at the staff's end and is fitted with a brass aperture disk. The horizon vane, with a 'bone', can slide along the staff. Drawing in pencil by A. Wijs, 1692. Courtesy, Museum Mr Simon van Gijn, Dordrecht, Netherlands.

doubt that, in North America too, the cross-staff was made and used into the eighteenth century.[20] Despite all these assurances, very little is known about the actual use of the instrument among seamen other than Dutch.

Through Davids we know about the predominance of the cross-staff over other contemporary instruments for the measurement of altitude in the Netherlands, in the period 1595 to 1650.[21] He investigated eighty-three estate inventories of Dutch seamen on the presence of mariner's astrolabes, cross-staffs and nautical quadrants. Forty-four contained one or more of these instruments, twenty-one had a cross-staff, and fifteen an astrolabe only. The rest showed combinations of a cross-staff with one or more of the other instruments.

The ships sailing on early Dutch voyages were provided with cross-staffs, nautical quadrants, circular and semi-circular astrolabes.[22] In 1655, the master and the three mates of every VOC ship each received a cross-staff. They were valued at 2 guilders and 11 *stuivers* per instrument, that is 10 *stuivers* more than the aforementioned Pieter Willemsz. had been paid by the VOC in 1620.[23] Each ship was also equipped with three mariner's astrolabes at 13 guilders each. After 1670 the provision of astrolabes was terminated as they were expensive, not very practical in use, and because their graduation was often inaccurate. Davids' research has further shown that disuse of the mariner's astrolabe at that time not only occurred in the VOC, but also in other branches of shipping. It was superseded by the cross-staff, of which increased numbers were found in estate inventories in the period 1676 to 1740.[24] Between 1731 and 1747, the year the octant was formally introduced on VOC ships,[25] the cross-staff was the only instrument for the measurement of altitudes. Not only was it being used, but also improved.

A report compiled by Douwes in 1749, and presented to the Amsterdam Admiralty, contained suggestions for the improvement of navigational methods and instruments, including the cross-staff. He wrote that there were errors in the diagrams used by cross-staff makers, resulting in erroneous instruments. Douwes reported having made a correct diagram, which he gave to the makers to use as an example. He also described an improvement for observing the sun backwards.

In 1762 the number of octants given to VOC ships was increased to two, one for the master and one for the chief mate. The other mates had to continue observing with cross-staffs. In 1780 three octants were provided and the number of cross-staffs was reduced to one. This was specifically for the cadets to learn how to observe with this instrument, as it was still in use on VOC ships sailing within the Indonesian Archipelago. After 1791 the cross-staff was no longer provided on VOC ships.[26]

The popularity of the cross-staff on VOC ships, was similar to that in the Dutch Admiralties. In 1747, the Rotterdam Admiralty ruled that its officers and mates were obliged to acquire their own cross-staffs, a rule which became effective in 1755.[27] According to the 1749 examination rules of the Amsterdam Admiralty, each mate seeking an appointment was obliged to demonstrate his skills, including observing with a cross-staff and an octant.

Disuse

From the above it has become clear that claims by historians that, by the end of the seventeenth century, the back-staff had replaced the cross-staff cannot be justified.[28] In fact, by 1700, the cross-staff had reached its highest level of perfection. During the next fifty years it was the most used instrument for altitude measurement by Dutch seamen and probably also by others.

After 1750, due to the acceptance of the octant, the use of the cross-staff receded. John Robertson, in 1754, when summing up the instruments used at sea for the measurement of altitudes, mentioned only the Davis quadrant and octant.[29] Somewhat reluctantly, he gave a brief description of the cross-staff, refraining from giving further details as it was difficult to use. New instruments for that purpose, he wrote, surpassing all others in accuracy and easy use, had recently been invented and come into general use. Obviously he was referring to the octant. In 1768 the Rotterdam teacher of navigation and examiner of the officers of the Rotterdam Admiralty, Antoni Struick, described the octant only because, as he wrote, the cross-staff should never be used because it was extremely inaccurate.[30] This is the last reference to the cross-staff found in the first edition of any navigational manual. Later authors, such as J. Hamilton Moore in 1772, did not mention the cross-staff at all, but described the octant only.[31] In the first Dutch standard work on the octant and sextant, published by the Amsterdam Admiralty Longitude Committee in 1788, it said that these were being used by most seamen, at least by all those who favoured an accurate observation.[32] Davids' research indicates that by this time the cross-staff was indeed no longer used by officers of large organizations such as the VOC and the Dutch Admiralties, but that this did not necessarily apply to other areas of shipping.[33]

A Comparison Between Surviving Examples of the Cross-Staff and Written Sources

In this chapter the three main elements of the surviving cross-staffs, the staff, the vanes, and the graduation, are described and compared with contemporary descriptions. The numbers refer to the Checklist.

The staff

In Spain, the staff was named *vara*. The Dutch named it *stok*, the French *fléche*, while the Italians called it a *bastone*.[1] Over the years, the general appearance of the surviving staffs remained the same. Most of them are made of ebony or lignum vitae and some of hornbeam; ivory is an exception. Apian, when referring to the astronomer's staff, mentioned oak or brass.[2] The latter material is consistent with the surviving astronomical cross-staffs.[3] Coignet and Zamorano mentioned wood, and Blundeville 'firm wood'.[4] The two oldest surviving staffs, dated ≤ 1596 and ≤ 1622 (Nos 1 & 2), are made of ebony, although the earliest mention in literature of that wood for staffs dates from 1643.[5] Besides ebony, Fournier named *Bresil*, known in English as redwood or braziletto (*Cæsaloina Chrita*), and pearwood[6], but only one staff made of pearwood was found (No.42). The VOC, in 1670, ruled that the cross-staffs provided to its ships should be made of ebony, as that wood did not warp.[7] Some years later Seller mentioned pearwood and boxwood, but no staff has been found to be made of the latter.[8]

If not damaged, the eye-end of all staffs is flat, often with a stamped instrument number or mark. No written reference to this was found. It is probably not the sequence number of the makers, but a batch number, for use by maker and user, in order to know which staff belonged to which vanes. As no two cross-staffs are exactly alike this would be useful, especially when several were in use simultaneously, such as on board the ships of the VOC. The numbers found are listed in Appendix IV.
The shape of all the staffs discovered is square; this is consistent with contemporary sources.[9] The sky-end of the staffs is usually shaped like a pyramid. No reference was found to this, but it can be seen in some contemporary illustrations.[10] It was probably shaped in this way so that the observer could distinguish the two ends of the staff by touch. This would be practical in the dusk or dark, when accidental reversing of the staff was likely. The oldest staff with a pyramid-shaped end is dated

Portrait of Antonij Kuijper (1762-1784), *élève* in navigation at the Fundatie van Renswoude in Utrecht. Kuijper sailed to Batavia as cadet on the *Brederode* of the Dutch East India Company, in 1783. In the foreground, the pyramid-shaped end of a staff with vane four is visible. Anonymous drawing. Courtesy, Fundatie van de Vrij-Vrouwe van Renswoude, Utrecht. Photograph Centraal Laboratorium, Amsterdam.

≤ 1676 (No.5), but the older staffs, worn or broken at the end, may have been shaped in the same manner.
The length of surviving staffs varies from about 900 to 1000 mm at the end of the seventeenth century, to about 750 mm at the end of the eighteenth century. These lengths have been listed in Appendix V, and, for comparison between theory and practice, the staff lengths given in contemporary navigational manuals are also listed there. This shows that, in theory, staffs were slightly longer than in practice. It also shows that there is very little variation in the lengths of Van Keulen staffs, in the period 1745 to 1760. This could mean that the firm had by then standardized the manufacture of cross-staffs.

The width of the surviving staffs varies from about 14 mm in the seventeenth century, to about 18 mm in the middle of the eighteenth century. It then decreases to about 16 mm towards the end of the century. These widths have been listed in Appendix VI. Contemporary written evidence on widths is scarce. The few authors who mention it are included in the Appendix.[11]
Some of the surviving staffs have been rounded off at the eye-end in order to fit the facial anatomy of the observer, and to correct for ocular parallax. Striking examples are the Dutch staff, ≤ 1596, and the Swedish, *c.* 1727 (Nos 1 & 39). Re-shaping of the eye-end is also visible on an early eighteenth century portrait of a shipmaster holding a cross-staff.
The ≤ 1596 staff (No.1), with two scales, has no side numbers. Side numbers on the ≤ 1622 staff (No. 2) were not found, as it is badly damaged. If not worn off or damaged, the other staffs, all with four scales, have the stamped digits 1, 2, 3 and 4 as side numbers at the eye-end, on each of the four sides. The staff by Thomas Tuttell (No.17) is the only one never to have had (visible) side numbers.
In general, side one was used to stamp in the maker's name or initials and the year of manufacture. This side, which is for the longest vane, provided the most blank space for stamps. There are, however, some exceptions to the rule. Although initially following it, after 1702 the Van Keulens invariably stamped their initials on side one and the year of manufacture on side two. Tuttell inscribed the staff with his name on side one and on side two. Side one of the anonymous American ? staffs (Nos 55 & 59) remained blank, while the year alone is found on side two.
On many staffs the maker's name or initials and year of manufacture have decorative stamps on either side. Often these decorations consist of a pattern of stars. In some cases, such as the No.1, the No.3, and the No.22 staffs, flowers were applied. The Van Keulens used either a single star, or a pattern of stars, set in a triangle or a square. These were stamped on both sides of their initials and on either side of the year of manufacture. The Van Keulen pattern, which changed over the years, is given in Appendix VII.
Many staffs have been damaged, some badly, when they were transformed for domestic purposes, e.g. for use as ells. At certain intervals, lines were carved in one or more of the staff's sides. In one case these intervals have been identified.[12] On some of the staffs, the ends are covered by metal caps. Others have holes drilled through near the eye-end, to attach a suspension string. Sometimes a hole for a screw, also for hanging the staff, was drilled in the flat eye-end, destroying any evidence of an instrument number. One staff was broken, reputedly when used to spank children, and two were used to straighten quilts.

The vanes

These were given a variety of names in the different countries. By the Spanish they were called *martilla*, *sualla* or *franja*.[13] The Fleming, Coignet, named them *transversariums*, and the French *transversaires*, *curseurs* or *marteaux*.[14] English authors called the vanes *transames*, *cursours*, *crosses*, *transums* or *cross-pieces*.[15] In Italian, they were known as *traverses* but, reputedly, seamen called them *castagnoles*.[16] The Swedish Admiral, Von Rosenfeldt, named

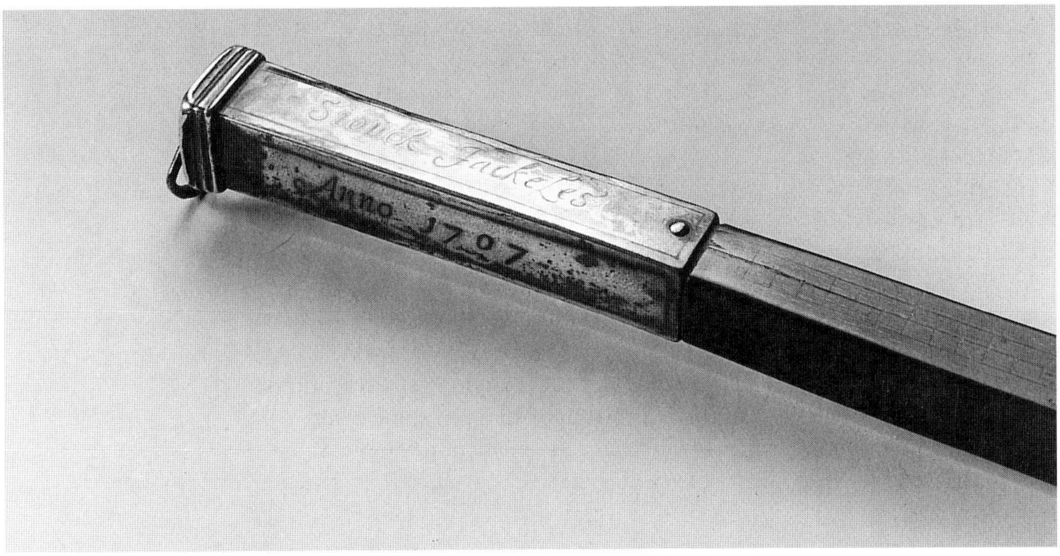

Silver cap on one end of the 1680 cross-staff (No.7). Courtesy, Fries Museum, Leeuwarden, Netherlands.

vanes *krusserna*[17], a translation of the common Dutch name *kruissen*, which means crosses. De Graaf was the only Dutch author who followed Coignet with *transversariums*.[18]
Original vanes are more rare than staffs. A likely reason for this is that, after use at sea, the staff could be adapted for a second life as a domestic object, while the vanes were of no further use. The oldest found vanes date from 1661 (No.3). Surviving examples are usually made from the wood of a fruit tree, often pear, and sometimes boxwood. Ivory (No.17) and ebony (No.38) are exceptions. Seller wrote that staff and vanes were usually made of the same wood, boxwood or pearwood.[19] Only with Nos 17 and 38 are vanes and staff made of the same material, but they are ivory and ebony. The first mention of pearwood, specifically for vanes, is by De Vries in 1702.[20] Also, in a 1777 sales catalogue of the firm of Van Keulen, pearwood vanes were advertized.[21]
The shape of all extant vanes is that of a stretched rectangle with a block in the centre. A square hole through vane and block allows for the staff to pass through. The 'spoon-shaped' vane, often seen on illustrations (e.g. the cover illustration of this book), has not been found.
De Vries was the first to give a comprehensive description of the clamping device that is found on most surviving vanes.[22] It consists of a small brass screw, which passes through the side of the block into the space through which the staff slides. A brass plate on the outside of the block fits over the entry hole of the screw, and prevents the wood from wearing out through frequent use. To prevent the screw from drilling into the staff, a second brass plate is wedged on the inside of the hole through which the staff slides. By tightening the screw when the observation is completed, this plate is pressed against the staff, holding the vane rigidly in position. According to De Vries, this improvement gave a more reliable reading. Despite the inner brass plate, the scales on some staffs show damage by an unprotected clamping screw, possibly because the plate was lost.

De Vries warned that vane and block should be made out of one piece of wood and not glued, as he claimed was frequently the case. Dampness would weaken the glue, rendering the vane useless.
Three of the oldest known vanes, probably English, (No.3), are constructed from two sections, glued together. The fourth vane, perhaps of later date, is made out of one piece of

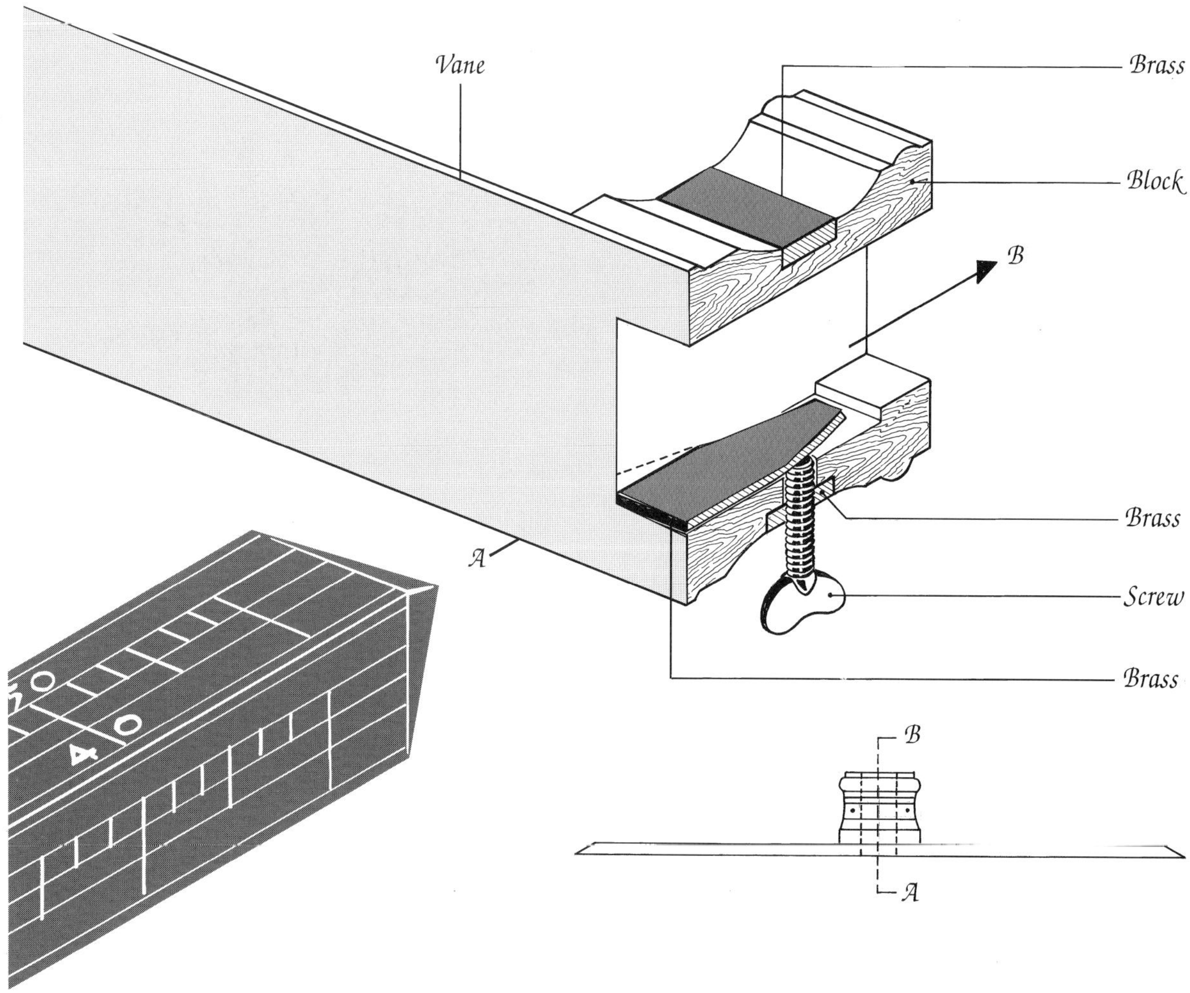

The construction of the clamping device of a vane.

wood. The Dutch vane (No.15) is also constructed from two sections, glued together. The English vanes of No.17 were constructed from two sections, kept together by thin metal bolts. The same construction applies to the *c.* 1700 vane (No.18) and the ≤ 1703 English vanes (No.22). The vanes of the cross-staff, part of a limewood carving, made in *c.* 1695 by Grinling Gibbons for the Admiralty Building in London, are also constructed from two sections, which, over the years, came apart.[23] As all later vanes are made out of one piece of wood, it could be concluded that not until around 1700 did this became normal practice. But, as most of the seventeenth century vanes are possibly English, this method of construction may be typical of that country.[24]

With the exception of one of the English ? 1661 vanes (No.3), and the Dutch 1699 (No.15), the seventeenth-century surviving vanes were made without a screw. It could, therefore, be concluded that the clamping screw was not in general use until around 1700. But, again, the absence of screws could be explained as typically English.

The boxwood horizon vane of an American ? 1666 cross-staff (No.4) with a nail, as decribed in 1677 by Millet Deschales, for the backward observation. Courtesy, Peabody & Essex Museum, Salem, Massachusetts.

Vane four, English, *c.* 1700, with wooden horizon (No.18). Courtesy, The Marquess of Bath, Longleat House, Warminster, Wiltshire.

Horizon vane with a 'bone', belonging to the 1760 cross-staff by Van Keulen (No.67). Courtesy, Universiteitsmuseum, Utrecht. Photograph Jac. P. Stolp.

The length of the vanes varies over the years. The surviving staffs show that, at the end of the seventeenth century, the longest vane could be well over 700 mm, far more than the 12 inches mentioned by authors like Coignet and Blundeville.[25] In the eighteenth century, the average length of this vane decreased to between about 500 and 650 mm. Only a few authors gave the lengths of the vanes, but several stated the ideal ratio of the lengths of the vanes and the staff. For many of the surviving staffs, the value is 1 : 2 : 4 : 6 : 9. The maximum variations found are 1 : (1½-3) : (3-6) : (4½-8) : (7-11½). These values agree with the ideal ratio given by contemporary authors. The ratios suggested in some of the manuals are:

1 : 4 : 8 (Coignet 1580; Fournier 1643)
1 : 2 : 4 (Blundeville 1594; Metius 1614)
1 : 2 : 4 : 8 (De Decker 1631; De Graaf 1658)
1 : 2 : 4 : 8 : 16 (Vooght 1684; Gietermaker 1742)
1 : 2 : 4 : 8 : 12 (De Vries 1702)

The only staff which, in this respect, differs from the others is the American ? (No.59), with a ratio of 1 : 2 : 7 : 9 : 9, meaning that staff and vane one have the same length.

The shortest vane was often called 'horizon' or 'horizon vane' because of its use when observing backwards. It was added to the cross-staff in the 1650s, probably shortly before 1655.[26] The oldest horizon vane dates from 1666 (No.4). A nail protruding from its side was to align the staff's end with the apparent horizon, when observing backwards. It was described in 1677 by Millet Deschales, who wrote how the shadow of the longer vane's upper edge should fall on the nail driven partly into the horizon vane, instead of using the wood or bone vane.[27] The latter element, in fact an extra vane, was probably added at the time the horizon was introduced, to improve the accuracy of the backward observation. It consisted of a flat piece of wood or bone, wedged into the horizon vane so that its straight edge was at a right angle to the horizon vane and intersected the heart of the staff. The upper edge of the bone vane was aligned with the apparent horizon and at the same time the shadow of the longer vane should fall on the bone. With this bone vane the cross-staff could be positioned far more accurately than with a plain horizon vane. The first mention of the extra vane originates from the VOC in 1655.[28] The oldest known *beentje* [little bone], as it was named in the 1742 edition of Gietermaker's manual[29], dates from 1661 (No.3). Only one wooden example survives, not detatchable but made out of one piece with the *c.* 1700 English vane (No.18). The one ivory example, also made of one piece with the horizon vane, dates from the same time (No.17). All other surviving *beentjes* are indeed made of bone and not permanently fixed to the vane. No reason was given for the use of bone, but it was probably preferred because the shadow of the longer vane contrasts better on white bone than on brown wood.
Another innovation of the 1650s was the brass aperture disk.[30] Using that, a more accurate backward observation was possible. It had to be slid on to the lower end of one of the longer vanes, leaving a slit between wood and brass. When looking through the slit and over the bone vane towards the apparent horizon, the cross-staff was accurately positioned. The first description of this device, in 1659, was by the Dutchman, Pietersz.[31] The fact that it was introduced and first used by the Dutch and that it resembled a shoe, explains why in English it was sometimes named Dutch shoe.[32] The oldest surviving example of such a disk dates from 1661 (No.3).
The aperture disk could also be fixed to the upper end of a vane, and hold a coloured glass to protect the eye from sunlight when observing facing the sun.[33] This glass, first mentioned by Bourne[34], was described by Colom in 1652 as being red or blue, and by Seller, in 1689, as red.[35] In 1777 the Van Keulens sold these glasses for six *stuivers*.[36] Only one such glass is known to survive. It is orange coloured and belongs to the 1711 cross-staff by Van Keulen

Front and back of the brass aperture disk fitted to the end of one of the vanes of the 1760 cross-staff by Van Keulen (No.67). Courtesy, Universiteitsmuseum, Utrecht. Photograph Jac. P. Stolp.

(No.28). Another probable application of the disk, for the backward observation, was introduced in 1749 by the Dutchman, Cornelis Douwes. Instead of the shadow of the large vane falling on the horizon vane, he wrote that the sunlight should fall on that vane through a slit, at the upper end of the large vane.[37] Although he did not describe how the slit was to be made, a brass aperture disk could very well serve for that purpose.

Moore and Colson described how the horizon vane could be omitted if the width of the next vane was equal to the length of the horizon vane.[38] Turned through 90°, its width could serve as horizon vane. No set of vanes has been found for which this is true. The width of the surviving vanes is generally *c.* 40 mm, which is about half the length of the average extant horizon vane.

Surviving vanes are not signed by the maker. Two have annotations in manuscript (Nos 15 & 48) and only one set is marked for the scale they belong to. They are English vanes (No.22), which are stamped with the digits 10, 30 and 90 for the respective scales they belong to. This is consistent with English authors, who name them the 'ten', 'thirty', 'sixty' and 'ninety' vanes.[39] In general, vanes have the same instrument number as the staff. It is usually stamped on the side of the clamping block, near the screw.

The scales

As described in the chapter Introducing the Cross-Staff, a third scale was proposed in 1580, and a scale for zenith distance in 1584. Around 1633, staffs were being made with combined scales for altitude and zenith distance, and around 1655, with the introduction of the fourth vane, all four sides of staffs were graduated.

The ≤ 1596 staff (No.1) has scales on two sides, one for altitude, the other for zenith distance. The $\frac{90}{0}$ marks of the scales, do not lie exactly opposite each other, but differ by 5 mm. This difference remains constant throughout the scales, enabling the observer to read the zenith distance from the staff, corrected for the sun's semi-diameter. The ≤ 1622 Spanish staff (No.2), has altitude scales only, on three sides. All other extant staffs are graduated on four sides for both altitude and zenith distance. This may not have been as common as is suggested by the surviving staffs. Colson, in 1685, still described staffs with a graduation for altitude only.[40]

According to contemporary authors, the range of graduation on staffs with three scales was for angles ≥ 30°, 10° to 30°, and ≤ 10°.[41] With the introduction of the fourth scale, the graduation was further spread to 30° to 90°, 20° to 60°, 10° to 30° and 3° to 10°.[42]

The division of the scales in degrees and minutes on surviving staffs is projected on four or six base lines which are finely cut along the length of the staff, starting at the $\frac{90}{0}$ mark and running to the pyramid-shaped end. The actual division consists of lines cut at right angles with the base lines, and is marked with stamped digits. On most staffs, these digits are stamped each ten degrees, for both altitude and zenith distance, e.g. $\frac{20}{70}$, $\frac{30}{60}$, and each five degrees in between with a stamped 5. Lines indicate further degrees and minutes. On staff No.1 each degree is marked by a stamped flower for zenith distance and a star for altitude. Each degree on staff No.3 is marked by a stamped flower. The degrees of staff No.56 are marked by a separate digit on sections of sides one, two, and three. The same is done on staff No.85 for the zenith distance scale of side three, between 75° and 78°.

In general, when holding the staff with side one upwards, the name or initials of the maker are at the left, between the eye-end and the beginning of the scale. When turning the staff away from the user, side two will appear, and then sides three and four. This sequence, in which the graduation on side one is for the longest vane and side two for the next, etc., is the same as given by Coignet in 1580.[43]
On some instruments, however, this sequence is different. On the 1661 staff (No.3) it is 1 2 4 3. On the French 1687 staff (No.10) the sequence is 4 3 2 1. Here, the name is on the right and the scales run from right to left. On the *c.* 1690, the 1691, and the 1731 staffs (Nos 11, 12 & 41) the sequence of the sides is also 4 3 2 1, but with the name at the left. This sequence could be of English origin. Moore numbered the sides starting with the shortest vane, and was followed in that by Colson and Seller.[44]

On some of the seventeenth-century staffs, (Nos 6, 7 and 9), the beginning of the scales is marked by a stamped sun for zenith distance, and a star for altitude (Pole star). The No.11 has a stamped sun and a half moon, while the No.3 has the letter D stamped at the beginning of the scale for zenith distance, which, perhaps, stands for 'distance'. Staff No.17 has the words 'Altitude' and 'Complement' engraved at the beginning of the respective scales on each of the four sides, and staff No.22 the abbreviation 'al'.
Early staffs were probably not divided in degrees and minutes. An illustration by Zamorano suggests a graduation in 60 minutes of arc.[45] A division in minutes mentioned by Haeyen was not found.[46] The oldest staffs (Nos 1 & 2) are divided in 30 minutes and 60 minutes re-

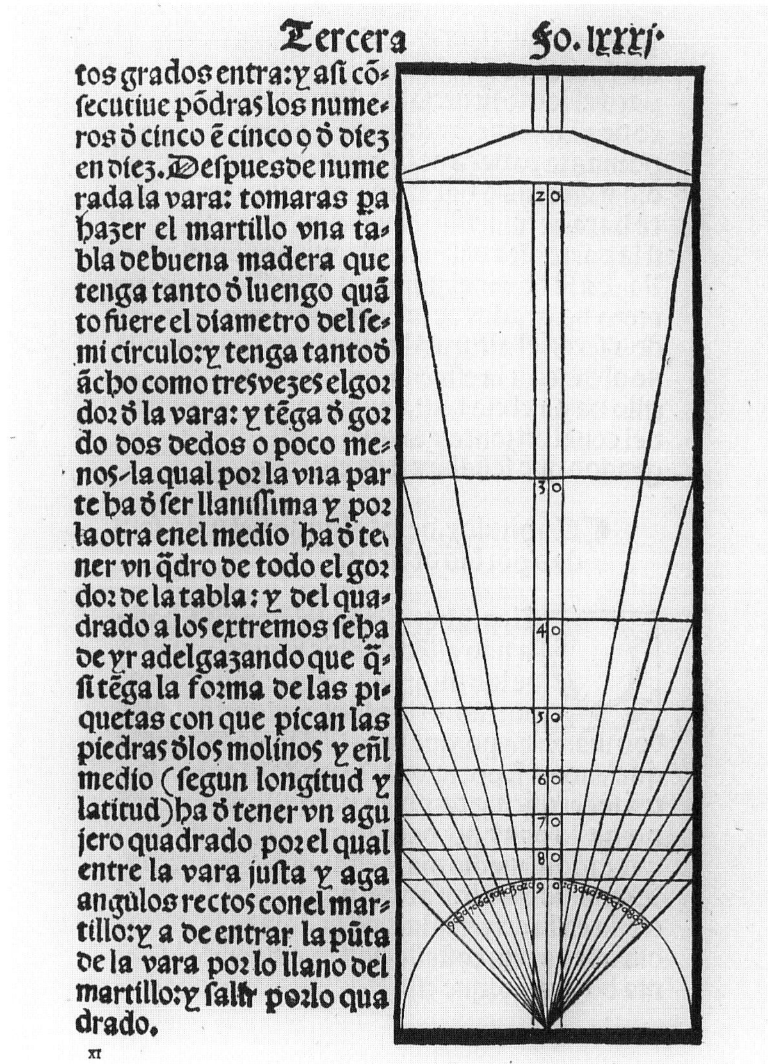
Tercera fo. lxxxi.
tos grados entra: y assi cõsecutiue pódras los numeros d cinco ẽ cinco o d diez en diez. Despues de numerada la vara: tomaras pa hazer el martillo vna tabla de buena madera que tenga tanto d luengo quãto fuere el diametro del semi circulo: y tenga tanto d ãcho como tres vezes el gordor d la vara: y tẽga d gordo dos dedos o poco menos / la qual por la vna parte ha d ser llanissima y por la otra en el medio ha d tener vn q̃dro de todo el gordor de la tabla: y del quadrado a los extremos se ha de yr adelgazando que q̃si tẽga la forma de las piquetas con que pican las piedras d los molinos y eñl medio (segun longitud y latitud) ha d tener vn agujero quadrado por el qual entre la vara justa y aga angulos rectos con el martillo: y a de entrar la pũta de la vara por lo llano del martillo: y salir por lo quadrado.
xi

The geometrical method of constructing the graduation, by Martin Cortes. From M. Cortes, *Breve Compendio*, Sevilla, 1551. Nederlands Scheepvaartmuseum, Amsterdam.

spectively, throughout the whole scales. The 1661 staff (No.3) is the earliest found to be divided to 10 minutes on the entire scale of side one. Side two also has a division in 10 minutes, but only for the range 30° to 60°, side three for the range 22° to 45° and side four for 7° to 30°. The 10 minute division is consistent with what De Vries wrote[47] in 1702, and with most other staffs surviving from the seventeenth century.

In the eighteenth century the divisions became finer, although a division in minutes, as suggested by Seller, was not found.[48] The oldest staff which is divided in 5 minutes of arc, on side four only, is dated *c.* 1700 (No.17). The next oldest is dated 1706 (No.23). On sides one and two of this staff, angles can be read to 5 minutes, and on sides three and four even to 2 minutes. Some staffs have several divisions, rising per side, e.g. the 1720 Hasebroek staff (No.35). On this, side four, between 50° and 80°, is divided in 60 minutes, between 40° and 50° in 30 minutes, between 30° and 40° in 15 minutes, between 10° and 20° in 5 minutes and between 7° and 10° in 2 minutes. Similar fine divisions are found on later staffs, in particular for sides three and four. This leads to the conclusion that there was a useful range.

The divisions found suggest that sides three and four, covering angles ranging from between 6° to 7° and 30° to 45°, were more commonly used than the other two, as they provided finer readings. The range agrees well with the 6° to 40° range that authors such as Harris, in 1730, advised as useful.[49] Because of the problem of blinking, the 40° limit is probably more realistic than the 50° to 60° which earlier authors advised.[50] Only one early author stated that, as the Pole star is not visible below 6°, there was little point in graduating a staff below that value.[51] Others proposed divisions down to 1°[52], but only one staff, an American ? (No.55), was found to be graduated to as little as 3°.

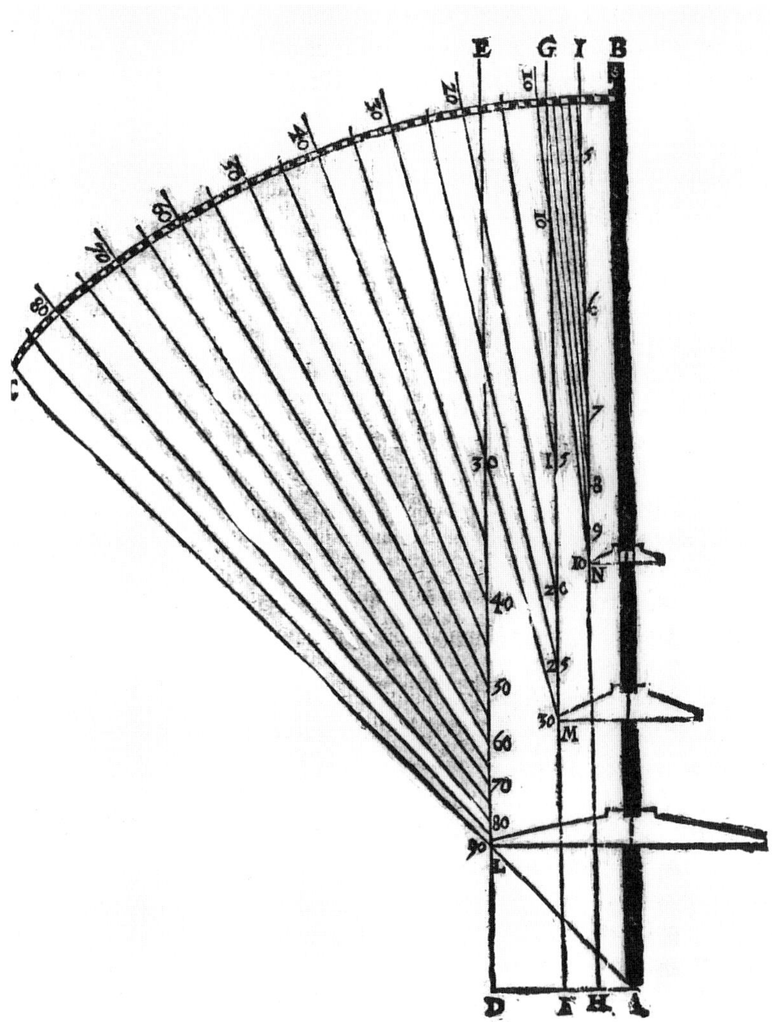

The geometrical method of constructing the graduation, by Michiel Coignet. From M. Coignet, *Nieuwe Onderwijsinghe*, Antwerp, 1580. Nederlands Scheepvaartmuseum, Amsterdam.

Conclusion

In the previous chapters the history of the construction, development, production, and diffusion of the cross-staff have been dealt with. The surviving examples are then described and the known cross-staff makers listed. This study has revealed two important new elements: the long period during which the instrument was used at sea, and Dutch predominance in its production.

It can be concluded beyond doubt that the cross-staff was widely used for a far longer period than has hitherto been accepted. It is not true, as both Cotter and Daumas state, that the cross-staff was replaced by the back-staff before the end of the seventeenth century. There is, as Davids has shown, enough evidence to indicate that by then the instrument had reached its highest level of perfection. Innovations, firmly implanted by 1700, made it the most accurate instrument for the measurement of altitudes at sea available at the time. The popularity of the cross-staff is reflected in the Dutch East India Company archives. The Amsterdam firm of Van Keulen alone sold 1148 cross-staffs to the VOC between 1731 and 1748, and these figures do not include sales to other branches of shipping. In fact, it is apparent that the Van Keulens were unable to meet the whole demand for cross-staffs in the first half of the eighteenth century, as in the same period others makers were also active in Amsterdam.

Around 1750, authors of navigational manuals ceased to publish descriptions of the cross-staff, even denouncing it in favour of the octant. The fact that this advice was followed is reflected in the numbers of surviving examples. The majority of the eighteenth-century instruments date from before 1750. Still, the production of cross-staffs continued after that year and for a considerable period too. Even the Van Keulens, who started producing octants in 1744, kept on making cross-staffs. The cross-staff remained in use at sea until the end of the century, although beyond doubt, to a far lesser extent than before.

It cannot easily be explained why so few non-Dutch examples have survived. Only one is French and possibly two are Spanish. No Portuguese cross-staff has been found, although it was used in association with the mariner's astrolabe, of which both Spanish and Portuguese examples survive.[1] Notwithstanding the fact that several British instrument makers made cross-staffs, and British seamen used them, only a few British examples survive. Until recently, only one cross-staff of established English origin was known, and that, made of ivory, is hardly typical of a seaman's navigating instrument.

An explanation for the discrepancy between the numbers of surviving Dutch and other staffs could be that the Dutch makers also served a large part of the foreign market. This hypothesis is supported by four arguments. Firstly, the fact that relatively few names of cross-staff makers from outside the Netherlands are known may indicate that there were not many of them. Secondly, explicit reference is made to Dutch cross-staffs by French and Swedish authors. Thirdly, many of the surviving Dutch cross-staffs now in German and Scandinavian collections originate from local seafaring families. Finally, what is known on the diffusion of nautical knowledge from the Netherlands to north-eastern Europe gives credence to this hypothesis. However, certain-

ty on these points can only be achieved when archival research on the production and diffusion of the cross-staff in seafaring countries has been done.
The fourth and most important explanation for the Dutch predominance is beyond doubt. The Dutch were responsible for major innovations in the instrument, all of which found support in other countries. Coignet, in 1580, was the first to propose the use of three scales, and Waghenaer in 1584 the first to provide a scale for zenith distance. Colom, in 1633, was the first to describe these two scales for each side of the staff and De Graaf, in 1658, was the first to mention scales on all four sides. The brass aperture disk, introduced by the Dutch in the 1650s, became known as the 'Dutch shoe'. About the same time, the Dutch introduced the wood or bone vane, to be mounted on the horizon vane. Both expedients improved the quality of backward observation. Finally, one of the methods of observing with the back to the sun was known in England as the 'Dutch fashion'. Experiments with a modern cross-staff have shown that this was superior to the normal method.

It appears that Dutch professional cross-staff makers, with the exception of the members of the Van Keulen family, were producing no other navigational instruments. Several, like Doncker, Loots, Mooy, and Haseboek, were chart makers. The Van Keulens, besides being authors and publishers of charts, maritime atlases and books, made a limited selection of nautical instruments, including compasses, back-staffs, octants, sextants and reflecting circles. The remaining Dutch cross-staff makers are not known to have produced other navigational instruments or sea-charts.

I hope that *The Cross-Staff*, as a study both of written sources and objects, will encourage similar research in the future. Besides the earlier proposal to carry out research into the production and diffusion of cross-staffs in seafaring countries, there are other factors that have not been dealt with extensively in this book and which I recommend for future study. One is the 'second life' of the staffs, for domestic use. Many staffs have carved lines and marks at intervals, probably representing feet or inches, but only in one case was the origin of these measures determined. Another element is the iconographical use of the cross-staff, in conjunction with the suggested biblical origin of the name, *baculus Jacob*. More frequently than other scientific instruments, the cross-staff was depicted on seventeenth and eighteenth-century vanitas paintings and allegories.
Finally, I should like to suggest a study of the astronomical and surveyors' staffs and the various staffs for navigation of which no examples survive. Such a study would provide a valuable contribution to the history of navigation and navigational instruments. I hope that, one day, these challenges may be taken up.

The Checklist

Introduction

The Checklist contains the numbered descriptions of 95 cross-staffs in chronological order of the date of manufacture. Most instruments are no longer complete. Often only the staff survives, sometimes transformed for domestic purposes, and in a few cases only one or more vanes survive.
Some cross-staffs have not been included. One of these is an unsigned and undated example in the National Technical Museum in Prague. Although that Museum claims it to be original, there is evidence to indicate that it is a copy.[1] Another cross-staff which has not been included, is the one referred to by E. Zinner, in the Lindauer Stadtmuseum.[2] During a recent check by museum personnel it was not found, nor any record of it ever having been in the museum. Other cross-staffs not included are two, one dated 1760, which were exhibited in Amsterdam in 1853 at a maritime exhibition.[3] They were displayed as historical objects by the Amsterdam instrument-making firms of J.M. Kleman & Zoon and the Weduwe G. Hulst van Keulen. The data on these is too flimsy to justify separate listing.[4] Finally, the fragment of a possible cross-staff in private ownership in Leeuwarden has been omitted. It would require research to establish if it is indeed part of a cross-staff.[5]
There exist a considerable number of modern reconstructions of cross-staffs, and copies of originals. These have not been included in the Checklist. Several museums use reconstructions for educational purposes or for display. The maritime museums in Helsingør, Denmark, in Stockholm, and in Oslo have a number of modern cross-staffs[6], several of which were made by Jens Kurk Jensen.[7] The National Maritime Museum, London owns a variety of replicas,[8] and Mystic Seaport Museum in Connecticut, U.S.A., has a cross-staff made locally in 1947.[9]
If known, any replica of an original in the Checklist is mentioned under *Remarks*. Harriet Wynter Ltd of London recently commissioned a series of modern cross-staffs, based on the two Dutch examples which were previously traded by the firm (Nos 14 & 48). They are clearly signed and dated, and cannot be mistaken for originals.

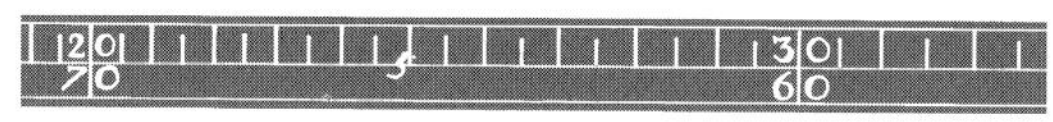

Section of a scale on a fragment of a possible cross-staff.

Two authors wrongly recorded original cross-staffs in museums in Oxford and Birmingham, England, and in Gdansk, Poland.[10] The authenticity of the cross-staff in the Civico Museo del Mare, Trieste, Italy,[11] could not be verified as the museum did not reply to my letter. A cross-staff reported in Rosenborg Castle, Copenhagen,[12] turned out to be a 117 cm long ceremonial staff made of wood covered with silver. It was commissioned in 1663 by King Frederik III of Denmark and Norway. The silver is engraved with astronomical data, such as epacts, the golden number, and a perpetual calendar. It also bears the misleading inscription 'Iacobsstaf'.

Inscriptions on the No.1.

Explanation

In the first line the sequence number of the checklist is given, followed by the maker's full name and the town where he worked, and the

staff's date. If the maker is not known, the name of the probable country of origin is given. Additional information on makers' names is provided in Appendix I. The date is the year of manufacture as stamped on the staff. When there is no such stamped-in year, an approximate dating according to other factors is entered, and argued under *Remarks*.
On the second line, the dimensions are first given. These are the total length of the staff, including the pyramid-shaped end, and the width, both in millimetres. If the staff has been shortened, or if it is a fragment, a plus sign indicates that it was longer. The dimensions are followed by the material of which the staff and vanes are made. NA indicates that this data was not available.
The *Instrument number* is the digit, letter or mark which is stamped in the flat eye-end of the staff and/or in the vanes.
Inscription records the maker's name, or initials, and the date of manufacture, and other stamps or decorations relating to the manufacture or ownership of the instrument.
Graduation gives the full scale for the altitude, for each side, starting with the longest vane. The *Range* is the graduation with the smallest division. This is, in fact, the useful range for that particular scale. The third figure is the maximum accuracy of reading found within the useful range. The presence or absence of a scale for zenith distance is mentioned under *Remarks*.
The lengths of the *Vanes* are given in millimetres in the sequence of the numbering of the sides, as with graduation. An added C means that the lengths of the vanes were calculated rather than measured. This value was found by multiplying by two the distance 90° to the eye-end. If known, length and width of the original vanes are entered under *Remarks*, and also information on the presence and quality of modern vanes.
Ratio is of the lengths of the vanes and the staff. It is, in fact, reversed, as the first digit represents vane four, the shortest.
The *Condition* of the instruments has three catagories:
Good: Staff in perfect condition and all original vanes present.
Fair: Staff undamaged, one or more vanes missing.
Poor: Staff (badly) damaged, one or more vanes missing.
Provenance gives the whereabouts of the instrument before it came to the present location.
Present location provides the name of the museum or collection where the staff is now kept, followed by, if known, its inventory number.
Literature gives the titles of works referring to the cross-staff in question, usually concerning provenance and present location.

The oldest known mariner's cross-staff, dated ≤ 1596 (No.1). The vanes do not belong to the staff. Courtesy, Rijksmuseum-Stichting, Amsterdam.

The list

1 Claes Andriesz. (Goutum) ≤ 1596

700+ mm; 13 x 13 mm; ebony

Instrument no. Not found
Inscription side 1
✶ DIT ✶ IS ✶ DIE ✶ NOERTSTERS ✶ SIIEDE ✶ GOVT ✶ OM ✶ DVECHT ✶
Inscription side 3
❀ DIT ❀ IS ❀ DIE ❀ SON ❀ SYDE ❀ CLAES ❀ ANDRIESSON ❀ GOVTVM ❀
Graduation sides
1 *full* 1° - 60°; ≥ 30'
3 *full* 90° - 40°; ≥ 30'
Vane
434 mm C
Condition Poor
Provenance Claes Andriesz. († 1597) probably was the maker and the owner. It was left by him on Nova Zembla after the 1596-1597 wintering of the Dutch, seeking the North-East passage. Retrieved by Charles Gardiner in 1876 and presented by him to the Dutch Government.
Present location Rijksmuseum, Amsterdam, The Netherlands, Nova Zembla collection, inv.NM7749a
Literature De Jonge, *Nova Zembla*; Cox, 'De jacobsstaf'; Brewington, 'Notes on the cross-staff', pp.188 & 190-191; Maddison, *Medieval Scientific Instruments*, pp.46-51; Crone, 'De vondst op Nova Zembla', pp.75-76; De Feyfer, 'De graadstok', p.17; *Focus Behaim Globus*, p.815, no.4.46; De Haan, 'Graadstokken', p.10
Remarks Graduated for zenith distance on side one, and for altitude on side three, the other two sides are blank. The distance from the eye-end to the beginnings of both scales differs by 5 mm. This difference, constant throughout the graduations, is the correction for the sun's semidiameter. The staff was probably longer; the end has been broken off. The eye-end has been rounded to fit the facial anatomy of the observer. Both accompanying vanes (187 and 52 mm) were, according to Crone, not found with the staff, and, as they have the wrong lengths, do not belong to it.
There is a plastic copy of this staff in the Gemeentelijk Museum 't Behouden Huys, West-Terschelling, The Netherlands, inv.461.

2 Not known (Spanish ?) ≤ 1622

867+ mm; 15 x 15 mm; ebony

Instrument no. Not found
Inscriptions Not found
Graduation sides
1 *full* 90°- 35°; ≥ 60'
2 ; ≥ 60'
3 ; ≥ 60'
Condition Poor
Provenance Recovered from the wreck of the Spanish ship *Nuestra Senora Atocha*, wrecked in 1622 off Key West, Florida, U.S.A.
Present location Treasure Salvors Inc., Key West, U.S.A., no.3013
Literature Mörzer Bruyns, 'Navigatie-instrumenten', p.276
Remarks No vanes. In extremely poor condition. Graduated for altitude on three sides. The graduation on sides two and three cannot be discerned.

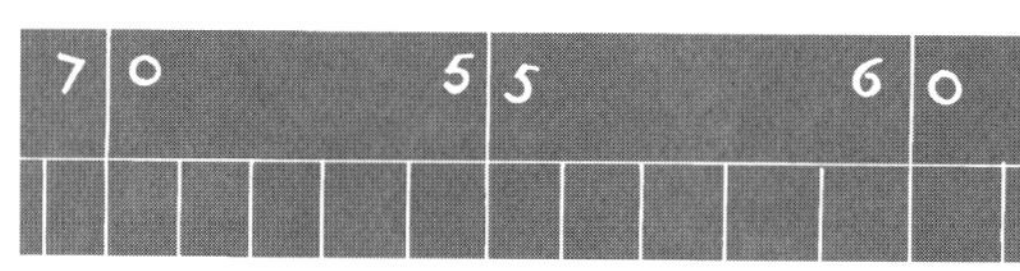

Section of a scale of the No.2.

3 Not known (English ?) 1661

1010 mm; $14^1/_2$ x $14^1/_2$ mm; fruitwood; *Vanes* pearwood
Instrument no. None
Inscription side 1
♠ · 1 · 6 · ♠ · 6 · 1 · ♠
Graduation sides
1 *full* 90° - 40°; *range* 90° - 40°; ≥ 10'
2 *full* 90° - 30°; *range* 60° - 30°; ≥ 10'
3 *full* 90° - 22°; *range* 45° - 22°; ≥ 10'
4 *full* 90° - 7°; *range* 30° - 7°; ≥ 10'
Vanes sides
1 724 mm C
2 518 mm C

3 298 mm C
4 90 mm C
Ratio of vane lengths & staff 1 : 3 : 6 : 8 : 11
Condition Fair
Provenance Wrangel or Brahe collection
Present location Skoklosters slott, Skokloster, Sweden, inv.10588
Literature Losman & Sigurdsson, 'Äldre vetenskapliga instrument'
Remarks With four original vanes. Vane four has part of a broken bone horizon. Vane two has a brass aperture disk with traces of black paint. The lengths of vanes one, two, and three are incorrect by -6 mm, +3 mm and -2 mm. They are constructed from two sections, glued together. All, except vane four, have clamping screws. Graduated for altitude, the complement is stamped on a parallel scale, at every 5°. On each side, this scale is marked at the beginning with a stamped letter D (distance ?). The sequence of the side numbers is 1 2 4 3. The staff is elaborately decorated with a stamped flower design.

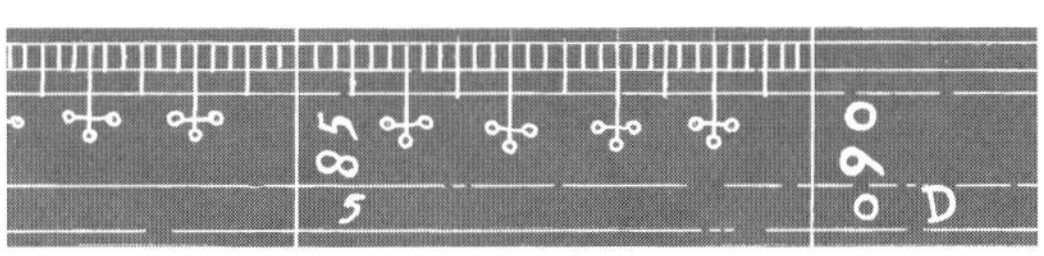

Beginning of the scale on side one of the No.3.

4 Not known (American ?) 1666
915 mm; 12 x 12 mm; ebony; *Vanes* boxwood

Instrument no. Not found
Inscription side 1
I I KAES H [or B] [Illegible]
1666
Inscription side 2
Edward Holyoke 1718
Graduation side
3 ≥ 10'
Condition Poor
Provenance Edward Holyoke (1689-1769), President of Harvard, and his son, Dr Edward Augustus Holyoke (1728-1829); remained in the family until placed in the Museum.
Present location Peabody & Essex Museum, Salem, Massachusetts, U.S.A., inv.M8727
Literature Brewington, 'Notes on the cross-staff', p.188; Brewington, 'Four American cross-staves', pp.139-140; Brewington, *The Peabody Museum*, p.6; Giordano, 'The Holyoke instruments', pp.38-39.
Remarks With a horizon vane without clamping screw, and one modern vane made in the Museum. All four sides were graduated, traces of which are now only partly visible on side three. Side two is divided 1-35 in quarter inches.

5 Not known (Swedish ?) ≤ 1676
89+ mm; 13 x 13 mm; ebony

Instrument no. Not found
Inscriptions Not found
Graduation sides
1 ; ≥ 10'
2 - 30°; ≥ 10'
3 ; ≥ 10'
4 ; ≥ 10'
Condition Poor
Provenance From the wreck of the Swedish warship, *Kronan*, sunk off Öland in the Baltic in 1676.
Present location Kalmar Läns Museum, Kalmar, Sweden, inv.KLM 18976 KR
Literature
Remarks No vanes. Fragment of the pyramid-shaped end section of a staff. Graduated for altitude and zenith distance.

6 Hendrik Pietersz. (Amsterdam) 1677
1005 mm; 14 x 14 mm; ebony

Instrument no. 7
Inscription side 1
* H · P *
1677
Graduation sides
1 *full* 90° - 40°; *range* 90° - 40°; ≥ 10'
2 *full* 90° - 29°; *range* 60° - 40°; ≥ 10'
3 *full* 90° - 17°; *range* 50° - 17°; ≥ 10'
4 *full* 90° - 6°; *range* 30° - 6°; ≥ 10'

Vanes sides
1 726 mm C
2 520 mm C
3 294 mm C
4 90 mm C
Ratio of vane lengths & staff 1 : 3 : 6 : 8 : 11
Condition Fair
Provenance Not known
Present location Privately owned, Amsterdam, The Netherlands
Literature
Remarks No vanes. Graduated for altitude and zenith distance, the beginnings of the graduations are marked respectively ⁕ and ✶. The end of the staff is pyramid-shaped.

7 Johan Sort ? (Amsterdam) 1680

960+ mm; 12 x 12 mm; ebony

Instrument no. Illegible
Inscription side 1

✶ ✶
✶ ✶ 16 SO 80 ✶ ✶
RT
✶ ✶

Graduation sides
1 *full* 90° - 18°; *range* 90° - 18°; ≥ 10'
2 *full* 90° - 32°; *range* 60° - 32°; ≥ 10'
3 *full* 90° - 18°; *range* 50° - 18°; ≥ 10'
4 *full* 90° - 6°; *range* 90° - 6°; ≥ 10'
Vanes sides
1 646 mm C
2 496 mm C
3 278 mm C
4 NA
Condition Poor
Provenance Ottema Kingma Stichting, Leeuwarden, The Netherlands
Present location Fries Museum, Leeuwarden, The Netherlands, inv.FM.6973
Literature Zijlstra, 'Jakobsstaven', p.4; De Haan, 'Graadstokken', p.11
Remarks No vanes. Graduated for altitude and zenith distance, the beginnings of graduations are marked respectively ⁕ and ✶. In very bad condition. Broken in one place, the ends are worn off, the staff was possibly longer. Both ends are fitted with silver caps, covering the beginning of the graduation on side four and possible side numbers. Therefore, the length of vane four and the ratio could not be calculated. The cap at the end of the staff, in fact a handle with suspension ring, is engraved with the Frisian names *Monte Freerx* and *Siouck Jackeles* and with *Anno 1707*. Monte Freerx († 1741), a widower, and farmer at Groot Jaarsma in Holwerd, married in 1704 Sjouckien Jackles from Westernijkerk. Sides two and four are fitted with silver studs and threads at intervals of 345; 23; 45; 175 and 520 mm.

Monogram and the beginning of the scale on side one of the No.7.

8 Not known (Dutch ?) 1682

1007+ mm; 14 x 14 mm; ebony

Instrument no. Not found
Inscription side 1

✶ /\ ✶
/✶ \ 1682 ✶ HD ✶

Graduation sides
1 *full* 90° - 41°; *range* 90° - 41°; ≥ 10'
2 *full* 90° - 30°; *range* 60° - 30°; ≥ 10'
3 *full* 90° - 18°; *range* 45° - 18°; ≥ 10'
4 *full* 90° - 5°; *range* 20° - 5°; ≥ 10'
Vanes sides
1 724 mm C
2 518 mm C
3 296 mm C
4 88 mm C
Ratio of vane lengths & staff 1 : 3 : $5\frac{1}{2}$: 8 : $11\frac{1}{2}$
Condition Poor
Provenance Found in a farmhouse in West-Friesland, The Netherlands; privately owned
Present location Nederlands Scheepvaartmuseum, Amsterdam, The Netherlands, on permanent loan, inv.RB.1283/H.134

Literature
Remarks With modern pearwood vanes of correct lengths, made in the Museum. Graduated for altitude and zenith distance. The inscription can be read as HD, HO or HC. There are additional marks carved on side four. Both ends of the staff are worn; possibly the staff originally was longer, consequently the vanes may have been longer than the replacements.

9 Johan Sort ? (Amsterdam) 1684

915+ mm; 13 x 13 mm; ebony

Instrument no. Not found
Inscription side 1

∗ ∗
∗ ∗ 16 SO 84 ∗ ∗
RT
∗ ∗

Graduation sides
1 *full* 90° - 47°; *range* 90° - 47°; ≥ 10'
2 *full* 90° - 31°; *range* 60° - 31°; ≥ 10'
3 *full* 90° - 17°; *range* ? - 17°; ≥ 10'
4 *full* 90° - 6°; *range* 30° - 6°; ≥ 10'
Vanes sides
1 691 mm C
2 499 mm C
3 270 mm C
4 89 mm C
Ratio of vane lengths & staff 1 : 3 : $5\frac{1}{2}$: 8 : 10
Condition Poor
Provenance The Fife Regional Council, Glenrothes, Scotland, U.K.
Present location National Museums of Scotland, Edinburgh, U.K., inv.1987.L9.2 or 1990.90
Literature Simpson, 'A Seventeenth Century Cross Staff'
Remarks No vanes. Graduated for altitude and zenith distance, the beginnings of graduations are marked respectively ⁕ and ∗. In very bad condition. Broken in several places and repaired. Both ends are fitted with brass caps, covering possible side numbers and any instrument number. The graduation on side three is almost completely erased. Was used as yardstick and has, on the brass caps, the weights & measures stamp for 1835.

10 Hamon (St Malo) 1687

915 mm; 13 x 13 mm; hornbeam ?

Instrument no. Not found
Inscription side 1

∗ T ∗
∗ HAMON · A · S · MALO · 1687 ∗
∗ ∗

Graduation sides
1 *full* 90° - 32°; *range* 60° - 32°; ≥ 15'
3 *full* 90° - 11°; *range* 30° - 11°; ≥ 15'
4 *full* 90° - 22°; *range* 30° - 22°; ≥ 15'
Vanes sides
1 510 mm
3 178 mm
4 346 mm
Ratio of vane lengths & staff 1 : 2 : ? : 3 : 5
Condition Fair
Provenance Thomas H. Court
Present location Science Museum, London, U.K., inv.1928-927, presented by T.H. Court in 1928
Literature Brewington, 'Notes on the cross-staff', p.188
Remarks With three modern vanes of correct lengths. Graduated for altitude. Only the number for side one is visible. Of the graduation on side two only vague traces are visible. The graduation runs from right to left. The sequence of the sides is 4 3 2 1.

Inscription on the No.10.

11 Not known (Dutch ?) *c.* 1690

830+ mm; 14 x 14 mm; ebony

Instrument no. 8
Inscription side 1
∗ L ∗ [Illegible]
Inscription side 3
1765
Inscription side 4
1784 [&] 1788
Graduation sides
1 *full* 90° - 42°; *range* 80° - 42°; ≤ 10'
2 *full* 90° - 27°; *range* 60° - 27°; ≤ 10'
3 *full* 90° - 12°; *range* 40° - 12°; ≤ 10'
4 *full* 90° - 7°; *range* 30° - 7°; ≤ 10'
Vanes sides
1 590 mm C
2 384 mm C
3 282 mm C
4 85 mm C
Ratio of vane lengths & staff 1 : 3 : 4 1/2 : 7 : 10
Condition Poor
Provenance Not known
Present location Gemeentelijk Museum 't Behouden Huys, West-Terschelling, The Netherlands, inv.628
Literature De Haan, 'Graadstokken', p.10
Remarks No vanes. In extremely poor condition. Graduated for altitude and zenith distance, the beginnings are marked respectively ☽ and ∗. Maker's name and side numbers have been deliberately erased. The inscriptions *1765*, *1784* and *1788* are of later date. The sequence of the side numbers is 4 3 2 1. The end of the staff has been re-shaped into a handle. There are marks carved at intervals of 45, 90, 180, 360 and 540 mm on side one. Side four has had a second graduation superimposed over the original.

12 Johannes van Keulen (Amsterdam) 1691

582 mm; 13 x 13 mm; ebony

Instrument no. H
Inscription side 1
∗ 16 ∗ 91 ∗
∗ ∗ ∗ ∗ ∗ ∗
∗ I ∗ V ∗ K ∗
Graduation sides
1 *full* 90° - 40°; *range* 80° - 40°; ≥ 10'
2 *full* 90° - 20°; *range* 50° - 20°; ≥ 10'
3 *full* 90° - 12°; *range* 30° - 12°; ≥ 10'
4 *full* 90° - 6°; *range* 20° - 6°; ≥ 10'
Vanes sides
1 414 mm C
2 194 mm C
3 118 mm C
4 60 mm C
Ratio of vane lengths & staff 1 : 2 : 3 : 7 : 10
Condition Fair
Provenance Found in the Museum's collection
Present location Provinciaal Overijssels Museum, Zwolle, The Netherlands
Literature Van Keulen, *In de Gekroonde Lootsman*, p.84
Remarks With modern vanes of correct lengths and two brass aperture disks. The clamping screws and the aperture disks are possibly original. The sequence of side numbers is 4 3 2 1. Graduated for altitude and zenith distance.

13 Johannes Loots (Amsterdam) 1696

770+ mm; 14 x 14 mm; ebony

Instrument no. I
Inscription side 1
∗ 16 ∗ 96 ∗
J ∗ Loots ∗
Graduation sides
1 *full* 90° - 44°; *range* 90° - 44°; ≥ 10'
2 *full* 90° - 30°; *range* 60° - 30°; ≥ 10'
3 *full* 90° - 13°; *range* 40° - 13°; ≥ 10'
4 *full* 90° - 7 ; *range* 20° - 7°; ≥ 10'

Vanes sides
1 601 mm C
2 396 mm C
3 166 mm C
4 86 mm C
Ratio of vane lengths & staff 1 : 2 : $4^{1}/_{2}$: 7 : 9
Condition Poor
Provenance Ottema Kingma Stichting, Leeuwarden, The Netherlands
Present location Fries Museum, Leeuwarden, The Netherlands, inv.FM.1956-285
Literature Zijlstra, 'Jakobsstaven', p.4
Remarks No vanes. Graduated for altitude and zenith distance. The end of the staff has been broken off.

14 Hendrik Doncker (Amsterdam) 1698
870+ mm; 14 x 14 mm; ebony

Instrument no. 4
Inscription side 1
∗ 1698 ∗
∗ H ∗ DONCKER ∗
Graduation sides
1 *full* 90° - 50°; *range* 90° - 50°; ≥ 10'
2 *full* 90° - 28°; *range* 60° - 28°; ≥ 10'
3 *full* 90° - 14°; *range* 40° - 14°; ≥ 10'
4 *full* 90° - 6°; *range* 20° - 6°; ≥ 10'
Vanes sides
1 690 mm C
2 424 mm C
4 90 mm C
Ratio of vane lengths & staff 1 : ? : 5 : $7^{1}/_{2}$: $9^{1}/_{2}$
Condition Poor
Provenance Private owners in Aberdeen, Scotland; Harriet Wynter Ltd., London
Present location Amsterdams Historisch Museum, Amsterdam, The Netherlands, inv.AMA.18264
Literature Harriet Wynter 1986, no.47; Wynter, 'Another 17th Century Cross-Staff', p.24
Remarks With modern vanes of correct lengths. Graduated for altitude and zenith distance. The graduation around 90° on side three has been erased, therefore the vane length could not be calculated. Side numbers have been erased. Carved-in marks on side three have been repaired. Probably slightly shortened at the end.

Inscriptions on the No.14.

15 Abraham Tade (Amsterdam) 1699
826 mm; 13 x 13 mm; ebony; *Vane* pearwood

Instrument no. H [in the staff only] D de Waay [in ink on the vane]
Inscription side 1
∗ 16 ∗ 99 ∗
∗ ∗
∗ A ∗ Tade ∗
Graduation sides
1 *full* 90° - 43°; *range* 90° - 43°; ≥ 10'
2 *full* 90° - 28°; *range* 60° - 28°; ≥ 10'
3 *full* 90° - 13°; *range* 40° - 13°; ≥ 10'
4 *full* 90° - 7°; *range* 20° - 7°; ≥ 10'
Vanes sides
1 612 mm C
2 388 mm C
3 176 mm C
4 86 mm C
Ratio of vane lengths & staff 1 : 2 : $4^{1}/_{2}$: 7 : $9^{1}/_{2}$
Condition Fair
Provenance
Present location Privately owned, Amsterdam, The Netherlands
Literature
Remarks With one original vane, 176 mm (width 38 mm). The vane is made in two parts, glued together. The screw is lost and there is no brass plate on the outside. The name D de Waay is probably that of an owner. Graduated for altitude and zenith distance. The end of the staff is pyramid-shaped. There are traces visible of an aperture disc (now missing) having been fitted to one end of the vane.

16 Johannes van Keulen (Amsterdam) 1700

660+ mm; 14 x 14 mm; ebony

Instrument no. Not found
Inscription side 1
✶ ✶1700✶ ✶
✶ ✶ ✶✶ ✶ ✶
✶ I✶V✶K ✶
Graduation sides
1 *full* 90° - 39°; *range* 80° - 39°; ≥ 10'
2 *full* 90° - 29°; *range* 70° - 29°; ≥ 10'
3 *full* 90° - 23°; *range* 40° - 23°; ≥ 10'
4 *full* 90° - 7°; *range* 20° - 7°; ≥ 10'
Condition Poor
Provenance H.J. de Feyfer, Terschelling, The Netherlands
Present location Nederlands Scheepvaartmuseum, Amsterdam, The Netherlands, inv.S.5116/H.109
Literature De Feyfer, 'De graadstok', p.12
Remarks No vanes. Graduated for altitude and zenith distance. The staff has been shortened at the eye-end. Consequently the side numbers have been erased and the vane lengths could not be calculated. There are marks carved at an interval of 350 mm on side one.

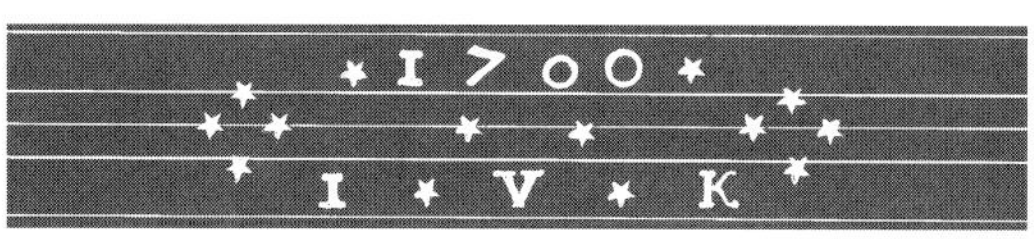

Inscriptions on the No.16.

17 Thomas Tuttell (London) *c.* 1700

796 mm; 14 x 14 mm; ivory; *Vanes* ivory

Instrument no. None
Inscription side 1
Thomas Tuttell Charing ✠ Londini fecit
Inscription side 2
Fait par Tho: Tuttell Ingenieur de sa MAIESTÉ BRITANIQUE pour les Instruments Mathematiques à LONDRES
Graduation sides
1 *full* 90° - 34°; *range* 90° - 34°; ≥ 10'
2 *full* 90° - 20°; *range* 60° - 20°; ≥ 10'
3 *full* 90° - 10°; *range* 30° - 10°; ≥ 10'
4 *full* 90° - 4°; *range* 11° - 4°; ≥ 5'
Vanes sides
1 506 mm
2 292 mm
3 136 mm
4 101 mm
Ratio of vane lengths & staff $1 : 1\frac{1}{2} : 3 : 5 : 8$
Condition Good
Provenance Purchased through Maggs Bros from the Rufford Abbey sale and presented to the National Maritime Museum, London
Present location National Maritime Museum, London, U.K., inv.S117/38-1370, presented by Sir James Caird in 1938
Literature Brewington, 'Notes on the cross-staff', p.188; *An Inventory*, I, 12-3; Holbrook, *Science Preserved*, p.164
Remarks No side numbers. Graduated for altitude and zenith distance. The end of the staff is pyramid-shaped. The beginnings of the scales on all four sides are marked *Altitude* and *Complement* respectively. Regarded as a presentation set. The vanes are made in two parts, held together by brass pins. Vane four has a fixed ivory horizon.

18 Not known (English ?) *c.* 1700

NA; 18 x 18 mm; *Vane* boxwood

Instrument no. 4
Vane side
4 47 mm
Condition Poor
Provenance Not known
Present location Longleat House, Longleat, Warminster, Wiltshire, U.K.

Signature of Thomas Tuttell on an ivory cross-staff, *c.* 1700 (No.17). Courtesy, Trustees of the National Maritime Museum, Greenwich, London.

Literature Heywood, 'A Brief Description', p.246; Holbrook, *Science Preserved*, p.148
Remarks The instrument has not been seen by the author. Vane four only, with a fixed wooden horizon. Made in two parts, attached by small brass tacks.

19 Johannes van Keulen (Amsterdam) 1701
872 mm; 14 x 14 mm; ebony

Instrument no. H, E or 7
Inscription side 1
* * 1701 * *
* * * * *
* I * V * K *
Graduation sides
1 *full* 90° - 42°; *range* 90° - 42°; ≥ 10'
2 *full* 90° - 28°; *range* 60° - 28°; ≥ 10'
3 *full* 90° - 12°; *range* 40° - 12°; ≥ 10'
4 *full* 90° - 5°; *range* 20° - 5°; ≥ 10'
Vanes sides
1 650 mm C
2 412 mm C
3 176 mm C
4 86 mm C
Ratio of vane lengths & staff 1 : 2 : 5 : 8 : 10
Condition Fair
Provenance Seafaring family on the island of Röm; Kgl. Navigationsschule, Altona, Germany
Present location Museum für Hamburgische Geschichte, Hamburg, Germany
Literature Schück, 'Der Jakobsstab', p.97; Schwindrazheim, 'Der Jakobsstab', no.1
Remarks With modern vanes of correct lengths. Graduated for altitude and zenith distance. According to Schück, the vanes were ordered by Director Engel of the *Kgl. Navigationsschule* and made by the Hamburg instrument maker, C. Plath. This is probably the staff described by Schück. The end of the staff is pyramid-shaped.

20 Johannes van Keulen (Amsterdam) 1702
755+ mm; 15 x 15 mm; ebony

Instrument no. Not found
Inscription side 1
* * *
* J * v * K *
* * *
Inscription side 2
* *
* 1702 *
* *
Graduation sides
1 *full* 90° - 39°; *range* 80° - 39°; ≥ 10'
2 *full* 90° - 26°; *range* 60° - 26°; ≥ 10'
3 *full* 90° - 13°; *range* 40° - 13°; ≥ 10'
4 *full* 90° - 6°; *range* 20° - 6°; ≥ 10'
Condition Poor
Provenance Dutch family collection
Present location Privately owned, Bussum, The Netherlands
Literature
Remarks No vanes. In extremely poor condition. Graduated for altitude and zenith distance. The staff has been shortened at the eye-end. Consequently the side numbers have been erased and the vane lengths could not be calculated. The staff is very badly damaged by carved marks on side two at intervals of 22, 44, 85, 170 and 335 mm. The end of the staff has been re-shaped into a handle.

21 Johannes Loots (Amsterdam) 1703
754 mm; 14 x 14 mm; ebony

Instrument no. Not found
Inscription side 1
* 1703 *
* J * Loots *
Graduation sides
1 *full* 90° - 42°; *range* 85° - 42°; ≥ 10'
2 *full* 90° - 28°; *range* 60° - 28°; ≥ 10'
3 *full* 90° - 14°; *range* 40° - 14°; ≥ 10'
4 *full* 90° - 7°; *range* 20° - 7°; ≥ 10'
Vanes sides
1 504 mm C
2 356 mm C

3 166 mm C
4 76 mm C
Ratio of vane lengths & staff 1 : 2 : $4^{1}/_{2}$: 7 : 10
Condition Poor
Provenance Donated to the Museum in 1934 by Ragnar Lörqvist of Gothenburg
Present location Sjöfartsmuseet i Göteborg, Gothenburg, Sweden, inv.SMG 3785
Literature
Remarks No vanes. Graduated for altitude and zenith distance. Although badly damaged, the staff does not appear to have been shortened. The side numbers have been erased.

22 Not known (English) ≤ 1703
837+ mm; 16 x 16 mm; wood; *Vanes* wood

Instrument no. Not found
Inscriptions Not found
Graduation sides
1 *full* 90° - 40°; *range* 75° - 40°; ≥ 10'
2 *full* 90° - 25°; *range* ? - 25°; ≥ 10'
3 *full* 90° - 15°; *range* ? - 15°; ≥ 10'
4 *full* 90° - 10°; *range* ? - 10°; ≥ 10'
Vanes sides
1 508 mm
2 302 mm
3 179 mm
4 118 mm
Ratio of vane lengths & staff 1 : $1^{1}/_{2}$: 3 : 5 : 7
Condition Poor
Provenance Recovered from the ship *Sterling Castle*, wrecked on the Goodwin Sands in 1703
Present location National Maritime Museum, London, U.K., inv.S.237/N82-36(1)
Literature
Remarks Badly damaged. Graduated for altitude and zenith distance. The beginning of the altitude scale is marked *al*, and the zenith distance scale possibly *co* for complement. The staff is incomplete and broken into six parts. Its end is a sawn-off pyramid. The four vanes are in several pieces. Staff and vanes were restored by the Museum in 1983. All vanes are made of two sections, held together by brass pins. The vanes are not fitted with clamping screws and are marked with a stamped *60*, *30* and *10* for the respective scales. It is possible that either one or three of the vanes do not belong to this staff.

The abbreviation *al* and decorative stamps at the beginning of a scale of the No.22.

23 Johannes van Keulen (Amsterdam) 1706
732 mm; $17^{1}/_{2}$ x $17^{1}/_{2}$ mm; ebony

Instrument no. 2
Inscription side 1
* *
* * J * v * K * *
* *
Inscription side 2
* *
* 1 * 7 * 0 * 6 *
* *
Graduation sides
1 *full* 90° - 39°; *range* 60° - 39°; ≥ 5'
2 *full* 90° - 27°; *range* 40° - 27°; ≥ 5'
3 *full* 90° - 14°; *range* 20° - 14°; ≥ 2'
4 *full* 90° - 7°; *range* 10° - 7°; ≥ 2'
Vanes sides
1 506 mm C
2 338 mm C
3 170 mm C
4 84 mm C
Ratio of vane lengths & staff 1 : 2 : 4 : 6 : 9
Condition Fair
Provenance Bought from a collector in the town of Roden, The Netherlands
Present location Privately owned, Westerbork, The Netherlands
Literature
Remarks No vanes. Graduated for altitude and zenith distance. The end of the staff is pyramid-shaped.

24 P. Boekenes? (Rotterdam) 1707
751 mm; 15 x 15 mm; ebony; *Vanes* pearwood

Instrument no. 4 H [in the staff]; 4 [in the vanes]
Inscription side 1
* 1707 *
* P * BOEKENES *
Graduation sides
1 *full* 90° - 38°; *range* 85° - 38°; ≥ 10'
2 *full* 90° - 28°; *range* 60° - 28°; ≥ 10'
3 *full* 90° - 13°; *range* 40° - 13°; ≥ 10'
4 *full* 90° - 7°; *range* 20° - 7°; ≥ 10'
Vanes sides
1 506 mm C
2 358 mm C
3 168 mm C
4 78 mm C
Ratio of vane lengths & staff 1 : 2 : 4½ : 6½ : 9½
Condition Fair
Provenance Bought in 1895 from private owners in the town of Arnis, Schleswig-Holstein, and presented to the Museum by Frau Margarethe Gaiser.
Present location Museum für Kunst und Gewerbe, Hamburg, Germany, inv.1895.212a-e
Literature Brinckman, *Museum für Kunst und Gewerbe*, pp.40-41; Schück, 'Der Jakobsstab', pp.94-95; Schwindrazheim, 'Der Jakobsstab', no.2; Dreier, *Winkelmessinstrumente*, pp.64 & 119, no.33; *Focus Behaim Globus*, pp.800-801, no.4.18; Eckhardt, *Wissenschaftliche Instrumente*
Remarks Three original vanes (vane three is missing) with correct lengths and one original brass aperture disk. Graduated for altitude and zenith distance. The end of the staff is pyramid-shaped. A Schleswig-Holstein foot is carved on the staff. The bone vane is missing.

25 Not known 1707

Condition Poor
Present location Not known, probably lost
Literature Schwindrazheim, 'Der Jakobsstab', no.35
Remarks The instrument has not been seen by the author. According to Schwindrazheim, the vanes were missing and the staff had been made to serve as an ell. In 1935 the staff was in the Heimatmuseum (now Stadtmuseum) Cuxhaven, Germany. A search through the Stadtmuseum in Cuxhaven, by the author and Dr U. Schnall in December 1987, did not reveal the instrument. The personnel of the Museum had no recollection of the staff, nor any documentation on it, so that it must be feared that it is lost.

26 Johannes van Keulen (Amsterdam) 1710
713+ mm; 14½ x 14½ mm; ebony

Instrument no. Not found
Inscription side 1
* J v K *
Inscription side 2
* 1710 *
Graduation sides
1 *full* 90° - 40°; *range* 65° - 40°; ≥ 5'
2 *full* 90° - 28°; *range* 40° - 28°; ≥ 5'
3 *full* 90° - 14°; *range* 25° - 14°; ≥ 5'
4 *full* 90° - 7°; *range* 20° - 7°; ≥ 5'
Vanes sides
1 508 mm C
2 340 mm C
3 170 mm C
4 86 mm C
Ratio of vane lengths & staff 1 : 2 : 4 : 6 : 8
Condition Poor
Provenance Miss A. Schweer of Leeuwarden, The Netherlands, descendant of a seafaring family.
Present location Eise Eisinga's Planetarium, Franeker, The Netherlands, donated by Miss Schweer in 1992
Literature De Haan, 'Graadstokken', p.10
Remarks With modern vanes of incorrect lengths. Graduated for altitude and zenith distance. Although the side numbers are still visible, the staff has been shortened at the eye-end. Consequently, the actual length of the vanes will have been longer. The end of the staff was probably also shortened. When latterly in private ownership, the staff was used to straighten quilts on beds.

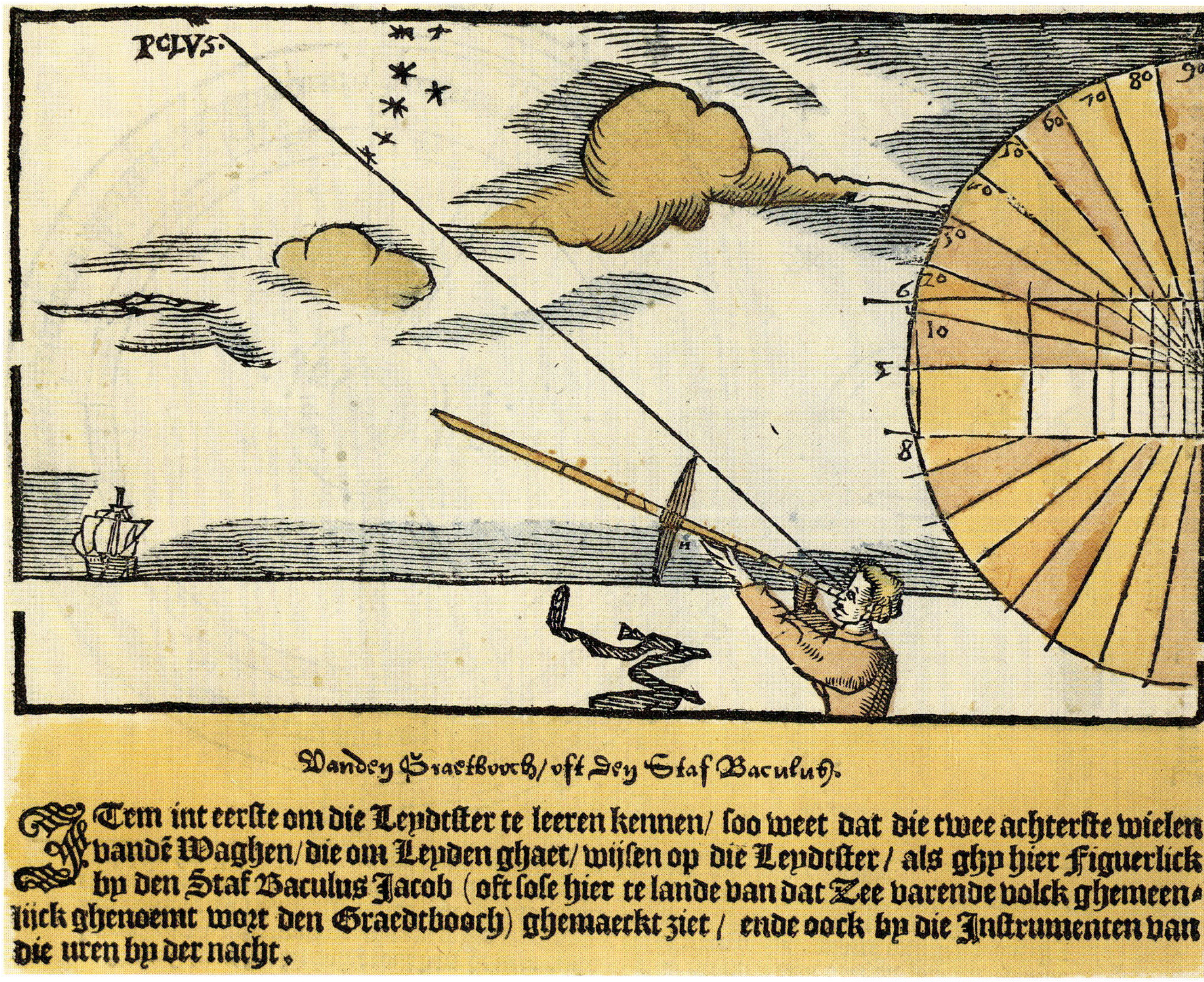

Observing the Pole star with a cross-staff. From G. Willemsen van Hollesloot, *Die Caerte vande Oost ende Westzee*, 1594
Nederlands Scheepvaartmuseum, Amsterdam

Portrait of an unknown shipmaster with cross-staff. The rounded eye-end of the staff, to compensate for ocular parallax, is visible. Oil on canvas, early eighteenth century. Courtesy, The Kendall Whaling Museum, Sharon, Massachusetts.

27 Johannes van Keulen (Amsterdam) *c.* 1710

738 mm; 14 x 14 mm; ebony; *Vanes* pearwood

Instrument no. Not found
Inscription side 1
⁎ J v K ⁎
G.D. BOM
Graduation sides
1 *full* 90° - 39°; *range* 70° - 39°; ≥ 10'
2 *full* 90° - 27°; *range* 60° - 27°; ≥ 10'
3 *full* 90° - 14°; *range* 30° - 14°; ≥ 10'
4 *full* 90° - 5°; *range* 20° - 5°; ≥ 10'
Vanes sides
1 508 mm C
2 342 mm C
3 170 mm C
4 84 mm C
Ratio of vane lengths & staff 1 : 2 : 4 : 6 : 9
Condition Fair
Provenance Sterrewacht, Utrecht, The Netherlands
Present location Universiteitsmuseum, Utrecht, The Netherlands, inv.UM 1884
Literature Ms UMU 'Catalogus der Sterrekundige Werktuigen', 1839, no.41 or 42; Ms UNU 'Catalogus der Sterrekundige Werktuigen', 1843, no.41 or 42; Ms UMU 'Catalogus der sterrekundige werktuigen' *c.* 1884, no.41 or 42; Ms UMU 'Catalogus der sterrekundige werktuigen', *c.* 1886, no.41(3); Ms UMU Inventaris voor 1937', no.41(3); Van Wijk, 'Nog eens de Jacobsstaf', p.54; Deiman, 'Het instrumentarium', p.187
Remarks The original vane (width 46 mm), badly damaged, for side three is present, and has the correct length (170 mm). Graduated for altitude and zenith distance. In *c.* 1886, there were two vanes with the staff. The end of the staff is pyramid-shaped. A suspension screw was fixed in the flat eye-end of the staff, erasing a possible instrument number. The name *G.D. Bom* is probably that of an owner.

28 Johannes van Keulen (Amsterdam) 1711

NA; NA

Instrument no. Not found
Inscriptions
[I V K 1711]
Condition Fair ?
Present location Heimatmuseum, Keitum, Sylt, Germany
Literature Schwindrazheim, 'Der Jakobsstab', no.4
Remarks The instrument has not been seen by the author. The staff, with the vanes for sides one and three, is kept in a decorated wooden box dated 1736. The box also contains a brass aperture disk, with orange coloured glass, to be fitted on the end of a vane for use as a sunlight filter. A copy of the box and the aperture disk were made in 1935 to fit the No.90.

29 Johannes Loots (Amsterdam) 1714

734 mm; 16 x 16 mm; ebony

Instrument no. 10
Inscription side 1
⁎ 1714 ⁎
⁎ J ⁎ Loots ⁎
Graduation sides
1 *full* 90° - 49°; *range* 80° - 49°; ≥ 10'
2 *full* 90° - 28°; *range* 60° - 28°; ≥ 10'
3 *full* 90° - 14°; *range* 40° - 14°; ≥ 10'
4 *full* 90° - 7°; *range* 20° - 7°; ≥ 10'
Vanes sides
1 508 mm C
2 358 mm C
3 168 mm C
4 80 mm C
Ratio of vane lengths & staff 1 : 2 : 4½ : 6 : 9
Condition Fair
Provenance Acquired from private owners in Utrecht, The Netherlands, *c.* 1968
Present location Privately owned, Zoelmond, The Netherlands
Literature
Remarks No vanes. Graduated for altitude and zenith distance. The end of the staff is pyramid-shaped.

Inscriptions on the No.29.

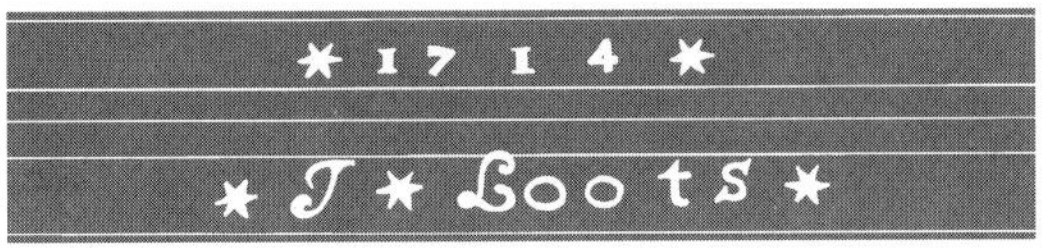

30 Not known (Dutch ?) 1714
827+ mm; 19 x 19 mm; ebony

Instrument no. Not found
Inscription side 1
* 1714 *
* * *
* F*G *
Graduation sides
1 *full* 90° - 42°; *range* 90° - 42°; ≥ 10'
2 *full* 90° - 28°; *range* 60° - 28°; ≥ 10'
3 *full* 90° - 12°; *range* 40° - 12°; ≥ 10'
4 *full* 90° - 5°; *range* 20° - 5°; ≥ 10'
Vanes sides
1 612 mm C
2 392 mm C
3 172 mm C
4 86 mm C
Ratio of vane lengths & staff 1 : 2 : $4^{1}/_{2}$: 7 : $9^{1}/_{2}$
Condition Poor
Provenance In *c.* 1810 this cross-staff belonged to a Captain Lammert Annes Kant of the flute, *De Liefde* (built 1773), sailing for Lammers, owners in Hindelopen, The Netherlands. Donated to the Museum in 1939 by Miss A. Hendriksma of Sneek.
Present location Fries Scheepvaart Museum, Sneek, The Netherlands, inv.J.032 (formerly FS-330)
Literature *Fries Scheepvaart Museum Gids*, p.7; De Haan, 'Graadstokken', p.9
Remarks No vanes. Graduated for altitude and zenith distance. The staff has been shortened at the eye-end, so that the calculated lengths of the vanes and ratio are possibly wrong.

31 Johannes Loots (Amsterdam) 1714

Inscriptions
[J. Loots 1714]
Condition Fair ?
Provenance From 1839 until at least 1937, belonging to the Sterrewacht, Utrecht, The Netherlands
Present location Not known
Literature Ms UMU 'Catalogus van Sterrekundige Werktuigen', 1839, no.41 or 42; Ms UMU 'Catalogus van Sterrekundige Werktuigen', 1843, no.41 or 42; Ms UMU 'Catalogus van sterrekundige werktuigen', *c.* 1884, no.41 or 42; Ms UMU 'Catalogus van sterrekundige werktuigen', *c.* 1886, no.41(2); Ms UMU 'Inventaris voor 1937', no.41(2); Van Wijk, 'Nog eens de Jacobsstaf', p.54
Remarks The instrument has not been seen by the author. In *c.* 1886, and in 1937, there were three vanes with the staff.

32 Johannes Loots (Amsterdam) 1716
828 mm; 16 x 16 mm; ebony; *Vanes* pearwood

Instrument no. 11 [in all parts]
Inscription side 1
* 1716 *
* J * Loots *
Graduation sides
1 *full* 90° - 41°; *range* 90° - 41°; ≥ 10'
2 *full* 90° - 28°; *range* 60° - 28°; ≥ 10'
3 *full* 90° - 12°; *range* 40° - 12°; ≥ 10'
4 *full* 90° - 7°; *range* 20° - 7°; ≥ 10'
Vanes sides
1 604 mm C
2 396 mm C
3 168 mm C
4 86 mm C
Ratio of vane lengths & staff 1 : 2 : $4^{1}/_{2}$: 7 : $9^{1}/_{2}$
Condition Good
Provenance See under remarks
Present location l'Observatoire Royal de Belgique, Brussels, Belgium
Literature Quetelet, *Annales*, p.17; *Catalogue Exposition*, no.E1; Rooseboom, *Bijdrage*, pp.40-43 & 93; *Inventaire*, 1, p.49

Ivory cross-staff by Thomas Tuttell, *c.* 1700, (No.17). Courtesy, Trustees of the National Maritime Museum, Greenwich, London.

Cross-staff by Johannes van Keulen, 1758, in contemporary box with decorations (No.66). Courtesy, Zuiderzeemuseum, Enkhuizen, Netherlands.

Remarks The original vanes have almost correct lengths (603, 395, 168 and 88 mm, width 37 mm). The bone horizon is missing. Graduated for altitude and zenith distance. The end of the staff is pyramid-shaped. The instrument belonged to the *Cabinet de Physique de la ville de Bruxelles* when that collection was presented to the Royal Observatory by order of the Minister of the Interior, in 1843. Director of this cabinet 1826 to 1833 was the Dutch instrument maker, J.H. Onderdewijngaart Canzius (1771-1838). The staff was not among the instruments of the Fundatie van Renswoude, which were acquired by Onderdewijngaart Canzius for the 'Cabinet'. The instrument was on loan to the Royal Museum in Brussels until 1984.

33 Johannes van Keulen (Amsterdam) 1717
744 mm; 15 x 15 mm; ebony

Instrument no. K
Inscription side 1
⋆ J v K ⋆
Inscription side 2
⋆ 1717 ⋆
Graduation sides
1 *full* 90° - 40°; *range* 80° - 40°; ≥ 10'
2 *full* 90° - 27°; *range* 60° - 27°; ≥ 10'
3 *full* 90° - 14°; *range* 35° - 14°; ≥ 10'
4 *full* 90° - 7°; *range* 20° - 7°; ≥ 10'
Vanes sides
1 506 mm C
2 338 mm C
3 170 mm C
4 84 mm C
Ratio of vane lengths & staff 1 : 2 : 4 : 6 : 9
Condition Poor
Provenance The Woudstra Collection, Ameland, The Netherlands
Present location Privately owned, Amsterdam, The Netherlands
Literature Zijlstra, 'Jakobsstaven', p.4
Remarks With modern vanes of correct lengths, made after No.48. Graduated for altitude and zenith distance. The end of the staff was pyramid-shaped, but now very much worn. Although sides two and three are badly damaged, all side numbers are still visible. Side one had the stylized crest of Ameland and side three *1739* carved in, both by the same hand, probably at the time when the staff was transformed into a stick for domestic measurement. At the sky end there is a hole through the stick for a suspension string. Side four has carved marks at intervals of 88; 177; 180; 357; 445 and 535 mm.

34 Not known (Norwegian ?) ≤ 1717
705 mm; 17 x 17 mm; NA; *Vanes* NA

Instrument no. Not found
Inscriptions Not found
Graduation sides
1 *full* 90° - 40°; ≥ 10'
2 *full* 90° - 30°; ≥ 10'
3 *full* 90° - 3°; ≥ 10'
4 *full* 90° - 3°; ≥ NA
Vane side
3 167 mm
Condition Poor
Provenance Retrieved from the wreck of the frigate, *Lossen*, which sank at the entrance of the Oslofjord in 1717
Present location Nørsk Sjofartsmuseum, Oslo, Norway
Literature Molaug & Scheen, *Fregatten 'Lossen'*, pp.137-138
Remarks The instrument has not been seen by the author. One original vane.

35 Jochem Hasebroek (Amsterdam) 1720
658 mm; 18 x 18 mm; ebony; *Vanes* pearwood?

Instrument no. H 3 [in the staff]; 3 [in the vanes]
Inscription side 1
1720
J ⋆ Hasebroek
Graduation sides
1 *full* 90° - 39°; *range* 60° - 39°; ≥ 5'
2 *full* 90° - 27°; *range* 40° - 27°; ≥ 5'

Inscriptions on the No.35.

Eye-end of the staff with side number 4, and flat eye-end showing instrument number, of the No.35.

Section of the scale of side four of the No.35.

Pyramid-shaped end of the No.35.

3 *full* 90° - 14°; *range* 30° - 14°; ≥ 5’
4 *full* 90° - 7°; *range* 10° - 7°; ≥ 2’
Vanes sides
1 456 mm C
2 306 mm C
3 152 mm C
4 74 mm C
Ratio of vane lengths & staff 1 : 2 : 4 : 6 : 9
Condition Good
Provenance The wreck of the 18th century Dutch galiot, *St Michael*, which sank in 1774 off the island of Borstö in Finland
Present location Museovirasto, Section for Maritime History, Helsinki, Finland, inv.0386:001
Literature Ericsson, ‘Borstö’, p.191; Mörzer Bruyns, ‘Navigatie-instrumenten’, p.276
Remarks The original vanes (width 41 mm) are damaged. The vane for side one is broken and only 360 mm long, for side two 304 mm, for side three 150 mm, and for side four 70 mm. Part of a broken bone horizon is attached to vane four. Graduated for altitude and zenith distance. The end of the staff is pyramid-shaped. Apart from the broken vane the instrument is in good condition.

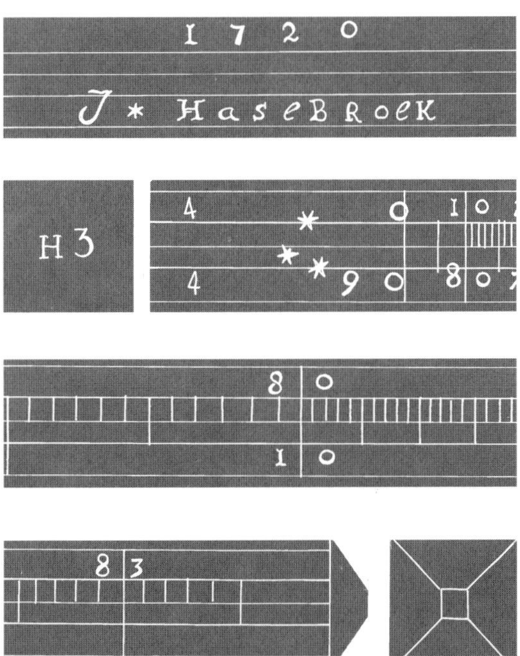

36 Johannes Loots (Amsterdam) *c.* 1720

722+ mm; 17 x 17 mm; ebony

Instrument no. Not found
Inscription side 1
*
* J * Loots *
Graduation sides
1 *full* 90° - 43°; *range* 85° - 43°; ≥ 10’
2 *full* 90° - 30°; *range* 60° - 30°; ≥ 10’
3 *full* 90° - 14°; *range* 40° - 14°; ≥ 10’
4 *full* 90° - 6°; *range* 20° - 6°; ≥ 10’
Vanes sides
1 506 mm C
2 356 mm C
3 NA
4 NA
Ratio of vane lengths NA
Condition Poor
Provenance Found in the Museum’s collection
Present location Museum für Kunst und Gewerbe, Hamburg, Germany, inv.1985.235
Literature Schück, ‘Der Jakobsstab’, p.97; Schwindrazheim, ‘Der Jakobsstab’, no.10; Eckhardt, *Wissenschaftliche Instrumente*
Remarks No vanes. Graduated for altitude and zenith distance. The staff is very badly damaged and now probably shorter than originally, so the vanes may have been longer than calculated. Due to damaged graduation, the lengths of the vanes for sides three and four could not be calculated.

37 Johannes van Keulen (Amsterdam) 1721

764+ mm; 15 x 15 mm; ebony

Instrument no. Illegible
Inscription side 1
* J * v * K *
Inscription side 2
* 1721 *
Graduation sides
1 *full* 90° - 39°; *range* 60° - 39°; ≥ 10’
2 *full* 90° - 26°; *range* 60° - 26°; ≥ 10’
3 *full* 90° - ?
4 *full* 90° - 7°; *range* 20° - 7°; ≥ 10’

Vanes sides
1 508 mm C
2 340 mm C
3 298 mm C
4 86 mm C
Ratio of vane lengths & staff 1 : 3½ : 4 : 6 : 9
Condition Poor
Provenance Donated to the Museum by A. Ecklund, Visby, Sweden in 1932
Present location Statens Sjöhistoriska Museet, Stockholm, Sweden, inv.S-1574
Literature Kaijzer, *Vraked vid Alvsnabben*, p.29
Remarks No vanes. Graduated for altitude and zenith distance. Both ends of the staff have probably been sawn off. Consequently, the vanes may have been longer than the calculated length. Except for the 90° mark, side three has been so badly rubbed that the graduation was completely erased.

38 Gerard van Keulen (Amsterdam) 1724
838 mm; 16 x 16 mm; ebony; *Vanes* ebony

Instrument no. 5 [in three of the vanes]
Inscription side 1
*   *
* *   * * *   * *
* * *   G * v * K   * * *
* *   *   * *
*   *
Inscription side 2
*   *
* *   *   *   * *
* * * 1 * * 7 ** 2 *** 4 * * *
* *   *   **   * *
*   *
Graduation sides
1 *full* 90° - 38°; *range* 60° - 38°; ≥ 5'
2 *full* 90° - 26°; *range* 40° - 26°; ≥ 5'
3 *full* 90° - 14°; *range* 30° - 14°; ≥ 5'
4 *full* 90° - 7°; *range* 20° - 7°; ≥ 5'
Vanes sides
1 560 mm C
2 374 mm C
3 188 mm C
4 94 mm C
Ratio of vane lengths & staff 1 : 2 : 4 : 6 : 9
Condition Fair
Provenance Sterrewacht, Utrecht, The Netherlands
Present location Universiteitsmuseum, Utrecht, The Netherlands, inv.UM 1883
Literature Ms UMU 'Catalogus van Sterrekundige Werktuigen', 1839, no.41 or 42; Ms UMU 'Catalogus van Sterrekundige Werktuigen', 1843, no.41 or 42; Ms UMU 'Catalogus van sterrekundige werktuigen', *c.* 1884, no.41 or 42; Ms UMU 'Catalogus van sterrekundige werktuigen', *c.* 1886, no.41(1); Ms UMU 'Inventaris voor 1937', no.41(1); Van Wijk, 'Nog eens de Jacobsstaf', p.54; Deiman, 'Het instrumentarium', p.187
Remarks With original vanes (559, 379, 187 and 94 mm). There is a bone horizon with vane four. Vane two is 5 mm too short; it possibly belonged to a different staff as, in *c.* 1886, there were only three vanes with this instrument. Graduated for altitude and zenith distance. The staff is damaged at the eye-end. The end of the staff is pyramid-shaped. A suspension screw was fixed in the flat eye-end of the staff, erasing a possible instrument number.

Inscription and decorative stamps on the No.38.

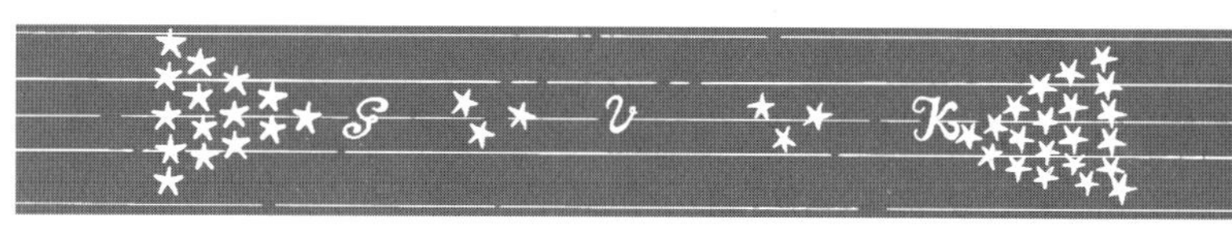

39 Not known (Swedish ?) *c.* 1727
740+ mm; 18 x 18 mm; ebony; *Vanes* NA

Instrument no. Not found
Inscriptions Not found
Graduation sides
1 ≥ 10'
2 ≥ 10'
Condition Poor
Provenance The Alvsnabben wreck, probably a galiot, found at Alvsnabben in the Stockholm Archipelago
Present location Statens Sjöhistoriska Museet, Stockholm, Sweden, find no. Asn 26 433 and

Asn 26 484
Literature Kaijzer, *Vraked vid Alvsnabben*, pp.29-31, 70-71, & 94
Remarks Part of a staff, broken into two pieces, with part of the clamping screw of one of the vanes. The graduation was so badly damaged that only traces could be found on sides one and two. The staff has been rounded at the eye-end to fit the facial anatomy of the observer.

40 Johannes Loots (Amsterdam) 1729
720 mm; NA

Instrument no. Not found
Inscription side 1
⋆ 1729 ⋆
⋆ J ⋆ Loots ⋆
Present location Lost
Literature Schwindrazheim, 'Der Jakobsstab', no.5
Remarks The instrument has not been seen by the author. Belonged to the Museum für Meereskunde, Berlin, Germany, lost during World War II. There is a copy in the Deutsches Museum, Munich, inv.21637, marked: *Nach original* MAX MARX *mechaniker* ***Berlin*** *1909. ⋆ 1729 ⋆ J ⋆ Loots ⋆*. In 1935 the original instrument had four vanes, of which one was newly made.

41 Not known (American ?) 1731
671 mm; 16 x 16 mm; ebony

Instrument no. 3
Inscription side 1
EN S TELMO
1731
Graduation sides
1 *full* 90° - 37°
2 *full* 90° - 27°
3 *full* 90° - 14°
4 *full* 90° - 7°
Vanes sides
1 428 mm C
2 300 mm C
3 150 mm C
4 74 mm C
Ratio of vane lengths & staff 1 : 2 : 4 : 6 : 9
Condition Fair
Provenance Saul Moscowitz, Historical Technology Inc., Marblehead, Massachusetts, U.S.A.
Present location The Time Museum, Rockford, Illinois, U.S.A., inv.1901
Literature
Remarks The lengths of the modern vanes are correct. Graduated for altitude and zenith distance. The end of the staff is pyramid-shaped. The sequence of the side numbering is the reverse of that on other staffs.

Inscriptions and decorative stamps on the No.41.

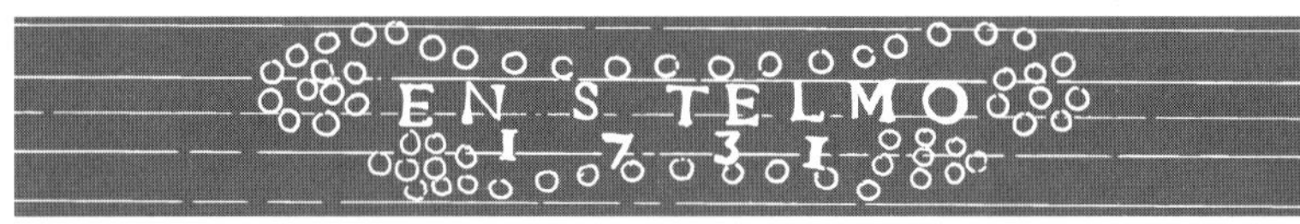

42 Not known (American) 1732
914 mm; 23 x 16 mm; pearwood ?; *Vanes* NA

Instrument no. None
Inscription side 1
E.H. ⋆ N.L. 1732.
Graduation side
1 *full* 75° - 37°; ≥ 60'
Condition Fair
Provenance Captain Louis F. Middlebrook USN, Louis F. Middlebrook Jr MD, Hartford, Connecticut, U.S.A., who donated it to the Museum in 1958
Present location Mystic Seaport Museum, Mystic, Connecticut, U.S.A., inv.58.1119
Literature
Remarks The accompanying vane (444 x 23 x 16 mm) is almost certainly not contemporary to the staff. Graduated, for altitude, on one side only. This instrument is probably not a (nautical) cross-staff.

Inscriptions and section of the scale on the No.42.

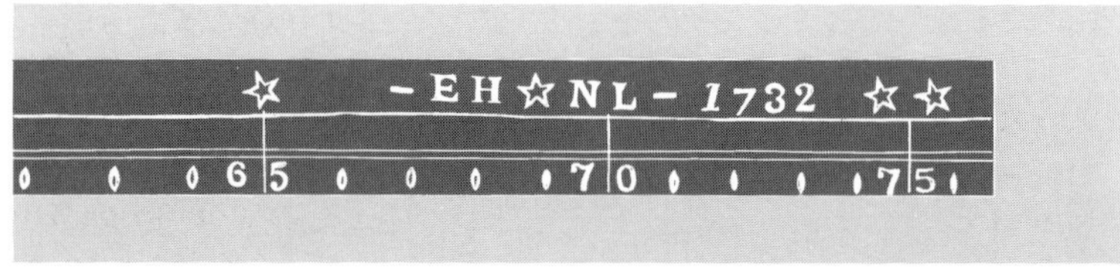

43 Johannes van Keulen (Amsterdam) 1733

731+ mm; 16 x 16 mm; ebony

Instrument no. Not found
Inscription side 1

*   *
* J * v * K *
*   *

Inscription side 2

*   *
* 1 * 7 * 3 * 3 *
*   *

Graduation sides
1 *full* 90° - 38°; *range* 60° - 38°; ≥ 5'
2 *full* 90° - 26°; *range* 40° - 26°; ≥ 5'
3 *full* 90° - 13°; *range* 30° - 13°; ≥ 5'
4 *full* 90° - 7°; *range* 20° - 7°; ≥ 5'
Condition Poor
Provenance Not known
Present location Museum für Hamburgische Geschichte, Hamburg, Germany, inv.AB434
Literature Schwindrazheim, 'Der Jakobsstab', no.6
Remarks No vanes. Graduated for altitude and zenith distance. The vane lengths could not be calculated as the staff has been shortened at the eye-end. The side numbers have been erased. The end of the staff is pyramid-shaped.

44 Jochem Hasebroek (Amsterdam) 1734

830 mm; 17 x 17 mm; ebony

Instrument no. HV[or A]
Inscription side 1
1734
J * Hasebroek
Graduation sides
1 *full* 90° - 40°; *range* 90° - 40°; ≥ 10'
2 *full* 90° - 27°; *range* 70° - 27°; ≥ 10'
3 *full* 90° - 12°; *range* 40° - 12°; ≥ 10'
4 *full* 90° - 7°; *range* 20° - 7°; ≥ 10'
Vanes sides
1 590 mm C
2 388 mm C
3 164 mm C
4 88 mm C
Ratio of vane lengths & staff 1 : 2 : $4^{1}/_{2}$: 7 : $9^{1}/_{2}$
Condition Fair
Provenance Found in the Museum's collection, acquired prior to 1894
Present location Museum für Kunst und Gewerbe, Hamburg, Germany, inv.1985.236
Literature Schück, 'Der Jakobsstab', p.97; Schwindrazheim, 'Der Jakobsstab', no.7; Eckhardt, *Wissenschaftliche Instrumente*
Remarks No vanes. Graduated for altitude and zenith distance.

45 Not known (German ?) *c.* 1735

830 mm; 19 x 19 mm; wood; *Vanes* wood

Instrument no. Not found
Inscriptions None
Graduation side
4 ≥ 10'
Vanes sides
2 389 mm
3 163 mm
Condition Fair ?
Provenance The Studiolo, a Ducal collection founded by Friedrich von Würtemberg (1593 - 1608)
Present location Würtembergisches Landesmuseum, Stuttgart, Germany, inv.KK 80
Literature Zinner, *Astronomische Instrumente*, p.210; *Christian IV and Europe*, p.255
Remarks The instrument has not been seen by the author. The two vanes (width 36 mm) are probably original; the vanes for sides one and four are missing. Zinner also saw vane four, probably with a bone horizon. Graduated for altitude and zenith distance. The end of the staff is pyramid-shaped. The staff has been dated from a photograph, according to its length and to that of the vanes. It resembles No.44. The dating in *Christian IV and Europe* must be doubted.

46 Not known (Dutch ?) ≤ 1735
NA; 17 x 17 mm; NA; *Vanes* fruitwood

Instrument no. Not found
Inscriptions Not found
Vane side
3 147+ mm
Condition Poor
Provenance Found by the British diver, Rex Cowan, in the wreck of the *T'Vliegend Hart*, a Dutch VOC ship which sank in the Wester Schelde in 1735
Present location Mr Rex Cowan, London, U.K., object T'Vliegend Hart no. 1905/90
Literature
Remarks The instrument has not been seen by the author. One very badly damaged vane for side three of a staff, only, with a fragment of a brass aperture disk attached. The brass clamping screw is missing.

47 Johannes van Keulen (Amsterdam) 1736
745 mm; 15 x 15 mm; ebony

Instrument no. Not found
Inscription side 1
[j v k]
Inscription side 2
∗ 1736 ∗
Graduation sides
1 *full* 90° - 38°
2 *full* 90° - 26°; ≥ 10'
3 *full* 90° - 14°
4 *full* 90° - 7°
Condition Fair ?
Provenance
Present location Privately owned, Vancouver, B.C. Canada
Literature Discovery 1978, p.13
Remarks With one modern vane. The instrument has not been seen by the author. Attempts by the curator of the Vancouver Maritime Museum to persuade the owner to supply additional details about the staff, failed. Graduated for altitude and zenith distance.

48 Not known (Dutch ?) *c.* 1740
NA; 15 x 15 mm; NA; *Vanes* pearwood

Instrument no. 3 [in all three]
Inscriptions I C L [in ink on all three]
Vanes sides
1 509 mm
2 339 mm
3 169 mm
Ratio of vane lengths [1] : 2 : 4 : 6
Condition Poor
Provenance Genootschap Felix Meritis, Amsterdam, The Netherlands
Present location Nederlands Scheepvaartmuseum, Amsterdam, The Netherlands, inv.B.25(5)/H.20
Literature Brewington, 'Notes on the cross-staff', p.188; Ms NSA 'Lijst van instrumenten', 1971
Remarks Three vanes only (width 40 mm), with a modern staff made for the museum in 1989 by W. van Lit. The clamping screws of vanes two and three are missing.

49 Johannes van Keulen (Amsterdam) 1740
858 mm; 15 x 15 mm; ebony

Instrument no. 18 or 81
Inscription side 1
∗ ∗
∗ J ∗ v ∗ K ∗
∗ ∗
Inscription side 2
∗ ∗
∗ 1740 ∗
∗ ∗
Graduation sides
1 *full* 90° - 37°; *range* 80° - 37°; ≥ 10'
2 *full* 90° - 25°; *range* 60° - 25°; ≥ 10'
3 *full* 90° - 13°; *range* 30° - 13°; ≥ 10'
4 *full* 90° - 7°; *range* 20° - 7°; ≥ 10'
Vanes sides
1 560 mm C
2 374 mm C
3 186 mm C
4 94 mm C

Ratio of vane lengths & staff 1 : 2 : 4 : 6 : 9
Condition Fair
Provenance Captain Solomon Davis of Eastern Long Island, N.Y., who sailed from New York, mid-eighteenth century. Bought by Alexander O. Vietor, New Haven, Connecticut, U.S.A. from a descendant of Davis. Bought from the Vietor estate by Harriet Wynter Ltd., London, U.K.
Present location The Lake St Louis Historical Society, Montreal, Canada
Literature Brewington, 'Four American cross-staves', p.139; *Harriet Wynter 1983*, no.7
Remarks With modern vanes of correct lengths. Graduated for altitude and zenith distance. The end of the staff is pyramid-shaped. Wynter wrongly dates the staff 173[6] or [0].

50 Johannes Loots (Amsterdam) 1741
719 mm; 16 x 16 mm; ebony; *Vanes* pearwood

Instrument no. Illegable
Inscription side 1
* 1741 *
* J * Loots *
Graduation sides
1 *full* 90° - 40°; *range* 85° - 40°; ≥ 10'
2 *full* 90° - 29°; *range* 60° - 29°; ≥ 10'
3 *full* 90° - 14°; *range* 40° - 14°; ≥ 10'
4 *full* 90° - 7°; *range* 20° - 7°; ≥ 10'
Vanes sides
1 508 mm C
2 356 mm C
3 166 mm C
4 78 mm C
Ratio of vane lengths & staff 1 : 2 : 4 1/2 : 7 : 9
Condition Fair
Provenance Belongs to the Museum's collection since 1913, deposited by the Göteborgs Museum
Present location Sjöfartsmuseet i Göteborg, Gothenburg, Sweden, inv.SMG 2389
Literature Navigationsskolorna 1841-1941, p.17
Remarks With two original vanes (width 38 mm) for the sides two (357 mm) and three (168 mm), broken and clamping screw missing, and a brass aperture disk. Graduated for altitude and zenith distance. The staff is damaged at the eye-end. A copy of this staff is in the Norsk Sjøfartsmuseum, Oslo, Norway.

51 Johannes Loots (Amsterdam) 1743
837 mm; 17 x 17 mm; ebony

Instrument no. Not found
Inscription side 1
* 1743 *
* J * Loots *
Inscription side 2
* A * C * BOER *
Graduation sides
1 *full* 90° - 40°; *range* 90° - 40°; ≥ 10'
2 *full* 90° - 27°; *range* 60° - 27°; ≥ 10'
3 *full* 90° - 12°; *range* 40° - 12°; ≥ 10'
4 *full* 90° - 7°; *range* 20° - 7°; ≥ 10'
Vanes sides
1 604 mm C
2 398 mm C
3 170 mm C
4 88 mm C
Ratio of vane lengths & staff 1 : 2 : 4 1/2 : 7 : 9 1/2
Condition Poor
Provenance A family in Venwoude, The Netherlands
Present location Westfries Museum, Hoorn, The Netherlands, inv.BR.84:87
Literature
Remarks No vanes. Graduated for altitude and zenith distance. The name stamped on side two appears to be of later date than 1743. The staff has been used for domestic measurement. A.C. Boer was probably an owner.

52 Not known (Dutch ?) ≤ 1743
139+ mm; 15 x 15 mm; ebony

Instrument no. Not found
Inscriptions Not found
Graduation sides
1 ≥ 5'
2 ≥ 5'
3 ≥ 5'
4 ≥ 5'

Condition Poor
Provenance From the wreck of the Dutch VOC ship, *Hollandia*, foundered off the Scilly Isles, 1743; Rex Cowan, London, U.K.
Present location Nederlands Scheepvaartmuseum, Amsterdam, The Netherlands, inv.S.4715(4)b/H.18
Literature Mörzer Bruyns, 'De navigatie-instrumenten', pp.45-46; *Hollandia Compendium*, p.457, no.2.9.10.1
Remarks No vanes. Fragment of a staff. Graduated for altitude and zenith distance. A brass aperture disk belonging to either this or the next cross-staff is in the Rijksmuseum, Amsterdam.

53 Jochem Hasebroek ? (Amsterdam) ≤ 1743

143+ mm; 14 x 14 mm; ebony

Instrument no. H
Inscriptions Not found
Graduation side
4 *full* 90° - ?; ≥ 60'
Vane side
4 74 mm C
Condition Poor
Provenance From the wreck of the Dutch VOC ship, *Hollandia*, foundered off the Scilly Isles, 1743; Rex Cowan, London, U.K.
Present location Nederlands Scheepvaartmuseum, Amsterdam, The Netherlands, inv.S.4715(4)a/H.18
Literature Mörzer Bruyns, 'De navigatie-instrumenten', pp.45-46; Mörzer Bruyns, 'Navigatie-instrumenten'; *Hollandia Compendium*, p.457, no. 2.9.10.1
Remarks No vanes. Fragment of a staff. Graduated for altitude and zenith distance. It is possible that this staff was made by Jochem Hasebroek, as No.35 by him has the same H as instrument number. A brass aperture disk belonging to either this or the previous cross-staff is in the Rijksmuseum, Amsterdam.

54 Johannes van Keulen (Amsterdam) 1745

744 mm; 16 x 16 mm; ebony; *Vanes* NA

Instrument no. 7
Inscription side 1
*   *
* J * v * K *
*   *
Inscription side 2
*   *
* 1745 *
*   *
Graduation sides
1 *full* 90° - 38°; *range* 80° - 38°; ≥ 10'
2 *full* 90° - 26°; *range* 60° - 26°; ≥ 10'
3 *full* 90° - 14°; *range* 35° - 14°; ≥ 10'
4 *full* 90° - 7°; *range* 20° - 7°; ≥ 10'
Vanes sides
1 510 mm C (from the copy)
2 340 mm C (from the copy)
3 168 mm C (from the copy)
4 84 mm C (from the copy)
Ratio of vane lengths & staff 1 : 2 : 4 : 6 : 9
Condition Fair ?
Provenance Not known
Present location Bergens Sjøfartsmuseum, Bergen, Norway, inv.B.Sj.302
Literature Horsburgh, 'The Cross-Staff'
Remarks The instrument has not been seen by the author. The three accompanying vanes, for sides one, two and three are possibly original. Graduated for altitude and zenith distance. The end of the staff is pyramid-shaped. The initials IR are painted on the staff, according to the Museum possibly by a previous owner. There is a good copy of this staff with three vanes in the National Museums of Scotland, Edinburgh, inv.1929.140, which was used for this description.

55 Not known (American ?) 1748

835 mm; 15 x 15 mm; ebony

Instrument no. 10
Inscription side 2
*   *
  * 1748 *
*   *
Graduation sides
1 *full* 90° - 38°; *range* 80° - 38°; ≥ 10'
2 *full* 90° - 26°; *range* 60° - 26°; ≥ 10'
3 *full* 90° - 13°; *range* 30° - 13°; ≥ 10'
4 *full* 90° - 3°; *range* 20° - 3°; ≥ 10'
Vanes sides
1 560 mm C
2 372 mm C
3 322 mm C
4 92 mm C
Ratio of vane lengths & staff 1 : $3\frac{1}{2}$: 4 : 6 : 9
Condition Fair
Provenance Captain Solomon Davis of Eastern Long Island, N.Y. who sailed from New York, mid-eighteenth century. Bought by Alexander O. Vietor, New Haven, Connecticut, U.S.A. from a descendent of Davis and presented to the Museum in 1952.
Present location Peabody & Essex Museum, Salem, Massachusetts, U.S.A., inv.M7799
Literature Brewington, 'Notes on the cross-staff', p.188; Brewington, 'Four American cross-staves', p.139; Brewington, *The Peabody Museum*, p.6
Remarks No vanes. Graduated for altitude and zenith distance. The end of the staff is pyramid-shaped.

56 Johannes van Keulen (Amsterdam) 1749

762 mm; 16 x 16 mm; ebony

Instrument no. 2 DO
Inscription side 1
*   *
  *J*v*K*
*   *
Inscription side 2
*   *
  1749
*   *
Graduation sides
1 *full* 90° - 39°; *range* 90° - 39°; ≥ 10'
2 *full* 90° - 29°; *range* 40° - 29°; ≥ 5'
3 *full* 90° - 12°; *range* 30° - 12°; ≥ 5'
4 *full* 90° - 6 ; *range* 20° - 6°; ≥ 5'
Vanes sides
1 534 mm C
2 374 mm C
3 158 mm C
4 74 mm C
Ratio of vane lengths & staff 1 : 2 : 5 : 7 : 10
Condition Poor
Provenance Not known
Present location Maritiem en Juttersmuseum, Texel, The Netherlands, inv.E.M. 111
Literature
Remarks No vanes. Graduated for altitude and zenith distance. Apart from two carved markings on side three, the staff is in reasonable condition. Digits have been stamped on parts of sides one, two and three for every degree for both altitude and zenith distance. The end of the staff is pyramid-shaped.

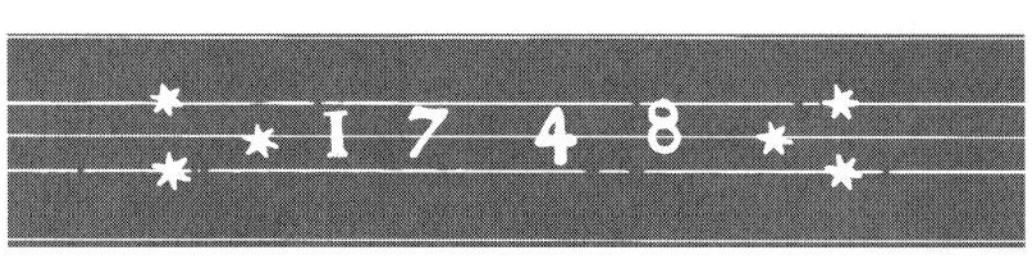

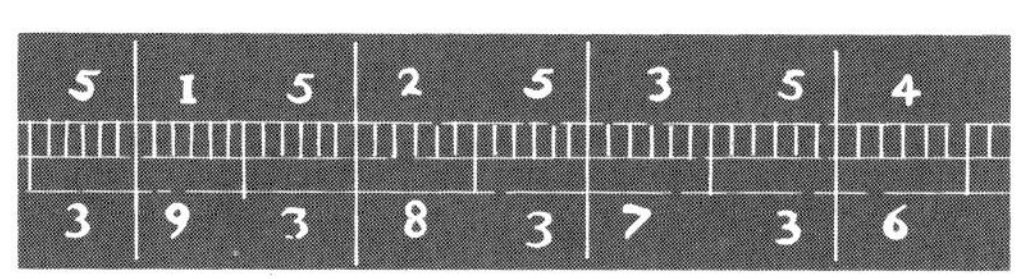

Left: Inscription on the No.55.

Right: Section of the scale on side two of the No.56.

57 Johannes van Keulen (Amsterdam) 1749

742 mm; 15 x 15 mm; ebony; *Vanes* pearwood

Instrument no. 6 or 9 [in all parts]
Inscription side 1
✶ ✶
✶ J✶v✶K ✶
✶ ✶
Inscription side 2
✶ ✶
✶ 1749 ✶
✶ ✶
Graduation sides
1 *full* 90° - 39°; *range* 80° - 39°; ≥ 10'
2 *full* 90° - 26°; *range* 60° - 26°; ≥ 10'
3 *full* 90° - 14°; *range* 35° - 14°; ≥ 10'
4 *full* 90° - 7°; *range* 20° - 7°; ≥ 10'
Vanes sides
1 510 mm C
2 340 mm C
3 170 mm C
4 84 mm C
Ratio of vane lengths & staff 1 : 2 : 4 : 6 : 9
Condition Good
Provenance Deutsche Seewarte, Hamburg, Germany
Present location Deutsches Hydrographisches Institut, Hamburg, Germany
Literature Schwindrazheim, 'Der Jakobsstab', no.8
Remarks Original vanes with modern brass fittings and clamping screws. The vanes have almost correct (509, 339, 169 and 85 mm) lengths and have been painted. The bone horizon is missing. Graduated for altitude and zenith distance.

58 Not known (Danish ?) *c.* 1750

567+ mm; 16 x 16 mm; ebony

Instrument no. Not found
Inscriptions Not found
Graduation sides
1 *full* 90° - 34°; *range* 85° - 39°; ≥ 10'
2 *full* 90° - 28°; *range* 60° - 28°; ≥ 10'
3 *full* ? - 14°; *range* 40° - 14°; ≥ 10'
4 *full* ? - 7°; *range* 20° - 7°; ≥ 10'
Condition Poor
Provenance Donated to the Museum in 1968 by Kai Rasmussen, Fyn, Denmark
Present location Handels- og Søfartsmuseet på Kronborg, Helsingør, Denmark, inv.122:68
Literature Thirslund, 'Samlingen af nautiske instrumenter', p.155
Remarks No vanes. A large portion of the staff has been sawn off at the eye end, hence the vane lengths could not be calculated. Graduated for altitude and zenith distance. The end of the staff is pyramid-shaped.

59 Not known (American ?) 1750

854 mm; 20 x 20 mm; ebony

Instrument no. 5
Inscription side 2
✶ ✶
✶ 1750 ✶
✶ ✶
Graduation sides
1 *full* 90° - 37°; *range* 80° - 37°; ≥ 10'
2 *full* 90° - 25°; *range* 60° - 25°; ≥ 10'
3 *full* 90° - 13°; *range* 30° - 13°; ≥ 10'
4 *full* 90° - 7°; *range* 20° - 7°; ≥ 10'
Vanes sides
1 860 mm C
2 676 mm C
3 188 mm C
4 94 mm C
Ratio of vane lengths & staff 1 : 2 : 7 : 9 : 9
Condition Fair
Provenance Captain Solomon Davis of Eastern Long Island, N.Y., who sailed from New York, mid-eighteenth century. Alexander O. Vietor, New Haven, Connecticut, U.S.A. obtained the staff from a descendent of Davis. Presented to the Museum by Vietor in 1953.
Present location Mystic Seaport Museum, Mystic, Connecticut, U.S.A., inv.53.44.58
Literature Brewington, 'Four American cross-staves', pp.138-140
Remarks The four modern vanes have incorrect lengths. Graduated for altitude and zenith distance. The end of the staff is pyramid-shaped.

60 Johannes van Keulen (Amsterdam) 1751
750+ mm; 16 x 16 mm; ebony

Instrument no. 2
Inscription side 1
∗ ∗
∗ I ∗ V ∗ K ∗
∗ ∗
Inscription side 2
∗ ∗
∗ 1751 ∗
∗ ∗
Graduation sides
1 *full* 90° - 38°; *range* 80° - 38°; ≥ 10'
2 *full* 90° - 26°; *range* 60° - 26°; ≥ 10'
3 *full* 90° - 13°; *range* 35° - 13°; ≥ 10'
4 *full* 90° - 7°; *range* 20° - 7°; ≥ 10'
Vanes sides
1 510 mm C
2 340 mm C
3 170 mm C
4 86 mm C
Ratio of vane lengths & staff 1 : 2 : 4 : 6 : 9
Condition Poor
Provenance
Present location Privately owned, West-Terschelling, The Netherlands
Literature
Remarks No vanes. Graduated for altitude and zenith distance. The side numbers have been sawn off. The pyramid-shaped end of the staff has partly been sawn off.

61 Jochem Hasebroek (Amsterdam) 1752
752+ mm; 16 x 16 mm; hornbeam ?

Instrument no. None
Inscription side 1
1752
J ∗ Hasebroek
Graduation sides
1 *full* 90° - ?; *range* 60° - ?; ≥ 10'
2 *full* 90° - ?; *range* 60° - ?; ≥ 10'
3 *full* 90° - ?; *range* 40° - ?; ≥ 10'
4 *full* 90° - ?; *range* 20° - ?; ≥ 10'
Vanes sides
1 508 mm C
2 356 mm C
3 170 mm C
Ratio of vane lengths & staff [1] : 2 : 4 : 6 : 9
Condition Poor
Provenance Not known
Present location Noordelijk Scheepvaartmuseum, Groningen, The Netherlands
Literature
Remarks No vanes. Graduated for altitude and zenith distance. The graduation of side four is so badly damaged around the 90° that is was not possible to calculate the vane length. The staff has no side numbers and may have been shortened. The staff has been adapted in order to be used to straighten quilts on beds.

62 Jochem Hasebroek (Amsterdam) 1753
843 mm; 16$^1/_2$ x 16$^1/_2$ mm; ebony

Instrument no. 1
Inscription side 1
1753
J ∗ Hasebroek
Graduation sides
1 *full* 90° - 40°; *range* 90° - 40°; ≥ 10'
2 *full* 90° - 27°; *range* 70° - 27°; ≥ 10'
3 *full* 90° - 12°; *range* 40° - 12°; ≥ 10'
4 *full* 90° - 7°; *range* 20° - 7°; ≥ 10'
Vanes sides
1 590 mm C
2 388 mm C
3 163 mm C
4 88 mm C
Ratio of vane lengths & staff 1 : 2 : 4 : 7 : 9$^1/_2$
Condition Fair
Provenance Bought, together with another cross-staff, in 1925 from the estate of schoolmaster Hendrik Klazes Bruning († 1925) in Dokkum, by bookseller and printer, Doederus Kamminga. Kamminga later became curator of the Admiraliteitshuis in Dokkum and donated the staff.
Present location Admiraliteitshuis, Dokkum, The Netherlands, inv.94
Literature De Feyfer, 'De graadstok'; De Haan, 'Graadstokken', p.9

Remarks The lengths of the modern vanes are not very accurate (595, 384, 165 and 87 mm), copied from the originals of No.66. Graduated for altitude and zenith distance. The end of the staff is pyramid-shaped. De Feyfer and De Haan wrongly date the instrument 1750.

63 Johannes van Keulen (Amsterdam) 1755

750 mm; 17 x 17 mm; ebony

Instrument no. 2
Inscription side 1
∗ ∗
∗ I ∗ V ∗ K ∗
∗ ∗
Inscription side 2
∗ ∗
∗ 1755 ∗
∗ ∗
Graduation sides
1 *full* 90° - 38°; *range* 80° - 38°; ≥ 10'
2 *full* 90° - 26°; *range* 60° - 26°; ≥ 10'
3 *full* 90° - 12°; *range* 35° - 12°; ≥ 10'
4 *full* 90° - 7°; *range* 20° - 7°; ≥ 10'
Vanes sides
1 508 mm C
2 340 mm C
3 170 mm C
4 *c.*86 mm C
Ratio of vane lengths & staff 1 : 2 : 4 : 6 : 9
Condition Poor
Provenance Not known
Present location Gemeentelijk Museum 't Behouden Huys, West-Terschelling, The Netherlands, inv.G.23
Literature De Haan, 'Graadstokken', p.10
Remarks No vanes. Graduated for altitude and zenith distance. The end of the staff is pyramid-shaped. The graduation on side four is badly damaged around 90°. The side numbers have been sawn off.

64 Johannes van Keulen (Amsterdam) 1756

743 mm; 17 x 17 mm; ebony

Instrument no. O
Inscription side 1
∗ ∗
∗ J ∗ v ∗ K ∗
∗ ∗
Inscription side 2
∗ ∗
∗ 1756 ∗
∗ ∗
Graduation sides
1 *full* 90° - 39°; *range* 80° - 39°; ≥ 10'
2 *full* 90° - 27°; *range* 60° - 27°; ≥ 10'
3 *full* 90° - 14°; *range* 35° - 14°; ≥ 10'
4 *full* 90° - 7°; *range* 20° - 7°; ≥ 10'
Vanes sides
1 510 mm C
2 340 mm C
3 170 mm C
4 72 mm C
Ratio of vane lengths & staff 1 : 2 : 4½ : 7 : 10
Condition Fair
Provenance Donated to the Museum in 1955 by Mrs Martine Christensen and Mrs Christine Ostergaard, in memory of their father, Ole Olesen Søndergård, a merchant. The staff had been found in an old seaman's chest which belonged to the donors' great-grandfather, Ole Nilsen Søndergård of Skarril, Denmark.
Present location Handels- og Søfartsmuseet på Kronborg, Helsingør, Denmark, inv.98:55, on loan to the Orlogsmuseet, Copenhagen, Denmark.
Literature Thirslund, 'Samlingen af nautiske instrumenter', p.155
Remarks With one modern vane (329 mm). Graduated for altitude and zenith distance. The end of the staff is pyramid-shaped.

65 Gerrit Hasebroek (Amsterdam) 1758
657 mm; 16 x 16 mm; ebony

Instrument no. None
Inscription side 1
1758
G ∗ Hasebroek
Graduation sides
1 *full* 90° - 39°; *range* 70° - 39°; ≥ 10'
2 *full* 90° - 25°; *range* 60° - 25°; ≥ 10'
3 *full* 90° - 13°; *range* 40° - 13°; ≥ 10'
4 *full* 90° - 7°; *range* 20° - 7°; ≥ 10'
Vanes sides
1 456 mm C
2 284 mm C
3 144 mm C
4 76 mm C
Ratio of vane lengths & staff 1 : 2 : 3½ : 6 : 8½
Condition Fair
Provenance Not known
Present location Dr.-Carl-Haeberlin-Friesen-Museum, Wyk auf Föhr, Germany, inv.S.0100
Literature
Remarks With three modern vanes (507, 338, and 168 mm). Graduated for altitude and zenith distance. The end of the staff is pyramid-shaped.

66 Johannes van Keulen (Amsterdam) 1758
749 mm; 18 x 18 mm; ebony; *Vanes* pearwood

Instrument no. 1 [in all parts]
Inscription side 1
∗ ∗
∗ I ∗ V ∗ K ∗
∗ ∗
Inscription side 2
∗ ∗
∗ 1758 ∗
∗ ∗
Graduation sides
1 *full* 90° - 38°; *range* 80° - 38°; ≥ 10'
2 *full* 90° - 26°; *range* 70° - 26°; ≥ 10'
3 *full* 90° - 13°; *range* 35° - 13°; ≥ 10'
4 *full* 90° - 7°; *range* 20° - 7°; ≥ 10'
Vanes sides
1 510 mm C
2 340 mm C
3 170 mm C
4 84 mm C
Ratio of vane lengths & staff 1 : 2 : 4 : 6 : 9
Condition Good
Provenance
Present location Zuiderzeemuseum, Enkhuizen, The Netherlands, inv.ZZM1953.B.930
Literature
Remarks The original vanes (width 41 mm) have almost correct lengths (508, 340, 168 and 84 mm). The bone horizon on vane four is broken. One brass aperture disk is present. The clamping screws of vanes two and three are missing. Graduated for altitude and zenith distance. The end of the staff is pyramid-shaped. The whole set is kept in an elaborately-decorated wooden box, possibly of later date.

67 Johannes van Keulen (Amsterdam) 1760
753 mm; 17 x 17 mm; ebony; *Vanes* pearwood

Instrument no. 12 [in all parts]
Inscription side 1
∗ ∗
∗ J ∗ v ∗ K ∗
∗ ∗
Inscription side 2
∗ ∗
∗ 1760 ∗
∗ ∗
Graduation sides
1 *full* 90° - 40°; *range* 80° - 40°; ≥ 10'
2 *full* 90° - 29°; *range* 60° - 29°; ≥ 10'
3 *full* 90° - 12°; *range* 35° - 12°; ≥ 10'
4 *full* 90° - 6°; *range* 20° - 6°; ≥ 10'
Vanes sides
1 526 mm
2 376 mm
3 151 mm
4 75 mm
Ratio of vane lengths & staff 1 : 2 : 5 : 7 : 10
Condition Good
Provenance Fundatie van Renswoude, The Netherlands
Present location Universiteitsmuseum,

Utrecht, The Netherlands, inv.UM.260, since 1932
Literature Van Wijk, 'Nog eens de Jacobsstaf'; Langenbach, *Onbekend talent*, pp.37-39
Remarks With original vanes (width 42 mm). Graduated for altitude and zenith distance. The end of the staff is pyramid-shaped. There is a bone horizon with vane four. There are two original brass aperture disks with the instrument. The name *I. Weteling*, in ink on vane one, is of Jan Weteling (*1747/48 † after 1776). He was a pupil at the *Fundatie van Renswoude* between 1761 and 1769. After 1763 he was educated in the art of navigation by the Fundatie's *instructor mathesis* Laurens Praalder who, from 1751 to 1761, was teacher in the art of navigation in Rotterdam and examiner of the mates of the Rotterdam Chamber of the VOC. Weteling joined the VOC and sailed as gunner's mate in 1765. He returned in 1767 and gave up the sea to become a school teacher. It would seem that this cross-staff was used at sea by Weteling and presented to the Fundatie after his career as seaman ended.

68 Johannes van Keulen (Amsterdam) 1762
743 mm; 17 x 17 mm; ebony

Instrument no. Not found
Inscription side 1
*   *
* J * v * K *
*   *
Inscription side 2
*   *
* 1762 *
*   *
Graduation sides
1 *full* 90° - 40°; *range* 80° - 40°; ≥ 10'
2 *full* 90° - 26°; *range* 60° - 26°; ≥ 10'
3 *full* 90° - 14°; *range* 40° - 14°; ≥ 10'
4 *full* 90° - 7°; *range* 20° - 7°; ≥ 10'
Vanes sides
1 510 mm C
2 338 mm C
3 168 mm C
4 84 mm C
Ratio of vane lengths & staff 1 : 2 : 4 : 6 : 9
Condition Fair
Provenance Private owners.
Present location Visserijmuseum, Vlaardingen, The Netherlands, inv.2587
Literature
Remarks No vanes. Graduated for altitude and zenith distance. The staff is damaged at the eye-end and also on the graduation on sides one (around 25°) and two (around 40°-45°). The end of the staff is pyramid-shaped.

69 Johannes van Keulen (Amsterdam) 1764
681 mm; 17 x 17 mm; ebony; *Vanes* NA

Instrument no. Not found
Inscriptions
[J v K 1764]
Graduation sides
1 *full* 90° - 38°; *range* 75° - 38°; ≥ 10'
2 *full* 90° - 26°; *range* 60° - 26°; ≥ 10'
3 *full* 90° - 13°; *range* 35° - 13°; ≥ 10'
4 *full* 90° - 7°; *range* 20° - 7°; ≥ 10'
Vanes sides
1 456 mm C
2 304 mm C
3 152 mm C
4 76 mm C
Ratio of vane lengths & staff 1 : 2 : 4 : 6 : 9
Condition Fair ?
Provenance Not known
Present location Tønder Museum, Tønder, Denmark, inv.TM 1383
Literature
Remarks The instrument has not been seen by the author. The data was provided by the Handels- og Søfartsmuseet på Kronborg in Helsingør, Denmark. Graduated for altitude and zenith distance.

70 Gerrit Hasebroek (Amsterdam) 1765
718 mm; 16 x 16 mm; ebony; *Vanes* pearwood

Instrument no. ɤ [in the staff only]
Inscription side 1
1765
G ∗ Hasebroek
Graduation sides
1 *full* 90° - 40°; *range* 85° - 40°; ≥ 10'
2 *full* 90° - 29°; *range* 60° - 29°; ≥ 10'
3 *full* 90° - 14°; *range* 40° - 14°; ≥ 10'
4 *full* 90° - 7°; *range* 20° - 7°; ≥ 10'
Vanes sides
1 510 mm C
2 358 mm C
3 168 mm C
4 78 mm C
Ratio of vane lengths & staff 1 : 2 : 4½ : 6 : 9
Condition Good
Provenance The widow of A.S. van der Sluis
Present location Streekmuseum 'Opsterland', Gorredijk, The Netherlands, inv.BL.5, on loan since 1962
Literature De Haan, 'Graadstokken', pp.9-10
Remarks With original vanes (width 39 mm) and a bone horizon on vane four. The vane lengths (508, 358, 167 and 78 mm) are almost correct. There are two vanes for side three. Graduated for altitude and zenith distance. The end of the staff is pyramid-shaped. Staff and vanes are kept in a wooden box, possibly of later date.

71 Johannes van Keulen (Amsterdam) 1765
737+ mm; 17 x 17 mm; lignum vitae ?

Instrument no. Not found
Inscription side 1
∗ ∗
∗ J ∗ v ∗ K ∗
∗ ∗
Inscription side 2
∗ ∗
∗ 1765 ∗
∗ ∗
Graduation sides
1 *full* 90° - 39°; *range* 80° - 39°; ≥ 10'
2 *full* 90° - 27°; *range* 60° - 27°; ≥ 10'
3 *full* 90° - 15°; *range* 45° - 15°; ≥ 10'
4 *full* 90° - 7°; *range* 20° - 7°; ≥ 10'
Vanes sides
1 506 mm C
2 334 mm C
3 290 mm C
4 82 mm C
Ratio of vane lengths & staff 1 : 3½ : 4 : 6 : 9
Condition Poor
Provenance Saul Moscowitz, Historical Technology Inc., Marblehead, Massachusetts, U.S.A.
Present location The Adler Planetarium, Chicago, Illinois, U.S.A. inv.A172
Literature
Remarks No vanes. Graduated for altitude and zenith distance. A hole has been pierced through the staff at the eye-end. The staff is worn, damaged at the eye-end, and possibly shortened. Consequently, the vanes may have been longer than calculated. The side numbers are worn off.

72 Johannes van Keulen (Amsterdam) 1765
658 mm; 18 x 18 mm; lignum vitae

Instrument no. 10
Inscription side 1
* *
* J * v * K *
* *
Inscription side 2
* *
* 1765 *
* *
Graduation sides
1 *full* 90° - 39°; *range* 80° - 39°; ≥ 10'
2 *full* 90° - 27°; *range* 60° - 27°; ≥ 10'
3 *full* 90 - 14°; *range* ? - 14°; ≥ 10'
4 *full* 90° - 7°; *range* 20° - 7°; ≥ 10'
Vanes sides
1 456 mm C
2 304 mm C
3 144 mm C
4 76 mm C
Ratio of vane lengths & staff 1 : 2 : 4 : 6 : 8½
Condition Fair
Provenance On loan to the Noordelijk Scheepvaartmuseum, Groningen, The Netherlands, from a private owner. Subsequently withdrawn and sold to a private owner at Kiel, Germany in 1988. Offered for auction at Sotheby's, London, May 1992, (estimate £ 18,000-24,000); not sold and returned to the owner, who donated it to the Museum.
Present location Kieler Stadt und Schiffahrtsmuseum, Kiel, Germany
Literature Marine Pictures, Nautical Works of Art and Medals, London, 20 May 1992. Lot.no.407, colour illustration; Baddeley, *Nautical Antiques*, p.52
Remarks With modern vanes, with almost correct lengths. There is a bone horizon, possibly original, with vane four. Graduated for altitude and zenith distance. The end of the staff is pyramid-shaped. Side three is damaged.

73 Not known (Dutch ?) 1765
667 mm; NA; *Vanes* NA

Inscriptions
[J D Z 1765]
Graduation sides
1 *full* 90° - 38°
2 *full* 90° - 26°
3 *full* 90° - 13°
4 *full* 90° - 6°
Vanes sides
1 446 mm C
2 304 mm C
3 152 mm C
4 76 mm C
Ratio of vane lengths & staff 1 : 2 : 4 : 6 : 9
Condition Fair-poor ?
Provenance Presented to Professor Niebuhr, Director of the Nautical School in Hamburg in 1874, by Dr Eduard Graffe of Hamburg
Present location Not known
Literature Wolf, *Astronomische Mittheilungen*, pp.23-24
Remarks The instrument has not been seen by the author. With four vanes (512, 440, 367 and 295 mm), not belonging to this staff. Graduated for altitude and zenith distance. The initials *J D Z* may be a misreading for J v K (Johannes van Keulen). According to Wolf, an exact copy of this staff was in the Observatory of Zurich in 1874.

74 Johannes van Keulen (Amsterdam) 1775
747 mm; 18 x 18 mm; ebony

Instrument no. 1.0
Inscription side 1
* *
* J * v * K *
* *
Inscription side 2
* *
* 1775 *
* *
Graduation sides
1 *full* 90° - 38°; *range* 80° - 38°; ≥ 10'
2 *full* 90° - 26°; *range* 60° - 26°; ≥ 10'

3 *full* 90° - 14°; *range* 35° - 14°; ≥ 10'
4 *full* 90° - 7°; *range* 20° - 7°; ≥ 10'
Vanes sides
1 510 mm C
2 339 mm C
3 170 mm C
4 85 mm C
Ratio of vane lengths & staff 1 : 2 : 4 : 6 : 9
Condition Fair
Provenance Bequest of J.A. Heuff, The Netherlands
Present location Museon, The Hague, The Netherlands, inv.45699
Literature Van Wijk, 'De Jacobsstaf', pp.40-42
Remarks With five modern vanes of correct lengths copied after the No.75. There are two vanes for side four, of which one has a modern bone horizon. Graduated for altitude and zenith distance. The end of the staff is pyramid-shaped.

75 Hendrik Mooy (Amsterdam) 1775

755 mm; 18 x 18 mm; ebony

Instrument no. 2
Inscription side 1
1775
* H * MOOY *
Graduation sides
1 *full* 90° - 38°; *range* 85° - 38°; ≥ 10'
2 *full* 90° - 28°; *range* 60° - 28°; ≥ 10'
3 *full* 90° - 13°; *range* 40° - 13°; ≥ 10'
4 *full* 90° - 6°; *range* 20° - 6°; ≥ 10'
Vanes sides
1 510 mm C
2 358 mm C
3 168 mm C
4 68 mm C
Ratio of vane lengths & staff 1 : 2 : 5 : 8 : 11
Condition Fair
Provenance Jacob Visser Jacobsz. Jnr of Schiermonnikoog, The Netherlands
Present location Maritiem Museum 'Prins Hendrik', Rotterdam, The Netherlands, inv.M.483
Literature Van Wijk, 'De Jacobsstaf', p.42; Brewington, 'Notes on the cross-staff', p.187; Jaspers, 'Over de constructie', p.499n
Remarks With modern vanes made by J. Visser of Schiermonnikoog. Graduated for altitude and zenith distance. The end of the staff is pyramid-shaped. The Museum had a replica of this staff made in its workshop.

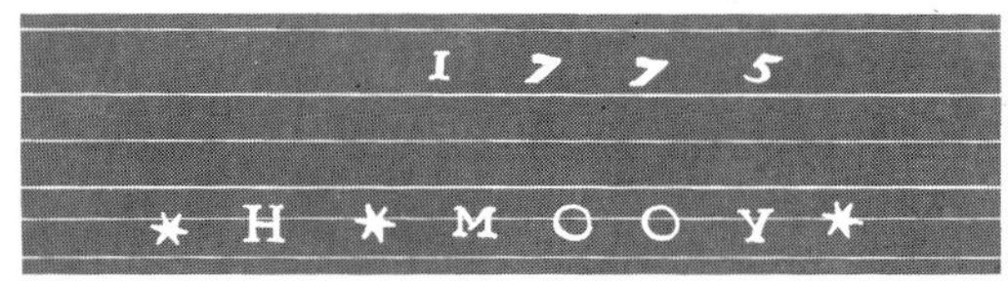

Inscriptions on the No.75.

76 Not known (Spanish ?) *c.* 1775

619 mm; 18 x 18 mm; ebony; *Vanes* NA

Instrument no. 4
Inscriptions None
Vanes sides
1 428 mm C
2 300 mm C
3 150 mm C
4 76 mm C
Ratio of vane lengths & staff 1 : 2 : 4 : 6 : 8
Condition Fair
Provenance
Present location Museo Naval de Madrid, Madrid, Spain, inv.I-289
Literature Franco, *Instrumentos Nauticos*, pp.70-72 & 198; Samso, *Instrumentos Astronomicos*
Remarks The instrument has not been seen by the author. There are three, possibly original vanes (429, 295 and 150 mm, width 39 mm) with the instrument. The staff has been dated by its dimensions.

77 Johannes van Keulen (Amsterdam) 1776
670 mm; 17 x 17 mm; ebony

Instrument no. 15
Inscription side 1
*   *
* J * v * K *
*   *
Inscription side 2
*   *
* 1776 *
*   *
Graduation sides
1 *full* 90° - 38°; *range* 75° - 38°; ≥ 10'
2 *full* 90° - 26°; *range* 60° - 26°; ≥ 10'
3 *full* 90° - 14°; *range* 40° - 14°; ≥ 10'
4 *full* 90° - 7°; *range* 20° - 7°; ≥ 10'
Vanes sides
1 454 mm C
2 304 mm C
3 152 mm C
4 76 mm C
Ratio of vane lengths & staff 1 : 2 : 4 : 6 : 9
Condition Fair
Provenance Purchased by the Musée *c.* 1830
Present location Musée de la Marine, Paris, France, inv.11.NA 5
Literature Brewington, 'Notes on the cross-staff', p.188
Remarks With two modern vanes (485 and 150 mm) and a bone horizon. Graduated for altitude and zenith distance. The end of the staff is pyramid-shaped. Brewington wrongly dated the instrument 1720.

78 Johannes van Keulen (Amsterdam) 1777
755 mm; 17 x 17 mm; ebony

Instrument no. 4
Inscription side 1
*   *
* J * v * K *
*   *
Inscription side 2
*   *
* 1777 *
*   *
Graduation sides
1 *full* 90° - 38°; *range* 80° - 38°; ≥ 10'
2 *full* 90° - 26°; *range* 70° - 26°; ≥ 10'
3 *full* 90° - 14°; *range* 45° - 14°; ≥ 10'
4 *full* 90° - 5°; *range* 20° - 5°; ≥ 10'
Vanes sides
1 510 mm C
2 338 mm C
3 168 mm C
4 86 mm C
Ratio of vane lengths & staff 1 : 2 : 4 : 6 : 9
Condition Fair
Provenance
Present location Noordelijk Scheepvaartmuseum, Groningen, The Netherlands, inv.5.523-1
Literature
Remarks Modern vanes. Graduated for altitude and zenith distance. The end of the staff is pyramid-shaped. The staff has been broken and repaired with brass, painted black.

79 Johannes van Keulen (Amsterdam) 1777
394+ mm; 18 x 18 mm; ebony

Instrument no. 8
Inscription side 1
*   *
* J * v * K *
*   *
Inscription side 2
*   *
* 1777 *
*   *

Graduation sides
1 *full* 90° - ?; *range* 80° - ?; ≥ 10'
2 *full* 90° - ?; *range* 70° - ?; ≥ 10'
3 *full* 90° - ?; *range* 40° - ?; ≥ 10'
4 *full* 90° - ?; *range* 20° - ?; ≥ 10'
Vanes sides
1 510 mm C
2 340 mm C
3 170 mm C
4 86 mm C
Ratio of vane lengths 1 : 2 : 4 : 6
Condition Poor
Provenance Not known
Present location Noordelijk Scheepvaartmuseum, Groningen, The Netherlands
Literature
Remarks Modern vanes. Graduated for altitude and zenith distance. A fragment of a staff only, lengthened and engraved in the Museum. Only the original engraving has been taken into account in this description.

80 Johannes van Keulen (Amsterdam) 1778
774 mm; 19 x 19 mm; lignum vitae; *Vanes* pearwood

Instrument no. 5 [in the staff]; 16 [in vane three]
Inscription side 1
* *
* J * v * K *
* *
Inscription side 2
* *
* 1778 *
* *
Graduation sides
1 *full* 90° - 38°; *range* 80° - 38°; ≥ 10'
2 *full* 90° - 26°; *range* 60° - 26°; ≥ 10'
3 *full* 90° - 13°; *range* 40° - 13°; ≥ 10'
4 *full* 90° - 7°; *range* 20° - 7°; ≥ 10'
Vanes sides
1 508 mm C
2 338 mm C
3 168 mm C
4 84 mm C
Ratio of vane lengths & staff 1 : 2 : 4 : 6 : 9
Condition Poor
Provenance Verifier of the instruments of the Royal Netherlands Navy
Present location Nederlands Scheepvaartmuseum, Amsterdam, The Netherlands, inv.B.195(2)/H.17
Literature Catalogus 1900, p.210, no.2009; Ms NSA 'Lijst van instrumenten', 1971
Remarks With two original vanes (338 and 168 mm, width 41 and 39 mm respectively). As vane two has no stamp, vane three is stamped 16 and both have different widths, they are probably from different staffs. Graduated for altitude and zenith distance. The end of the staff is pyramid-shaped. A hole has been drilled through the staff near the eye-end, probably for an suspension string. Probably the same instrument as mentioned in *Catalogus 1900*, however, wrongly dated 1774.

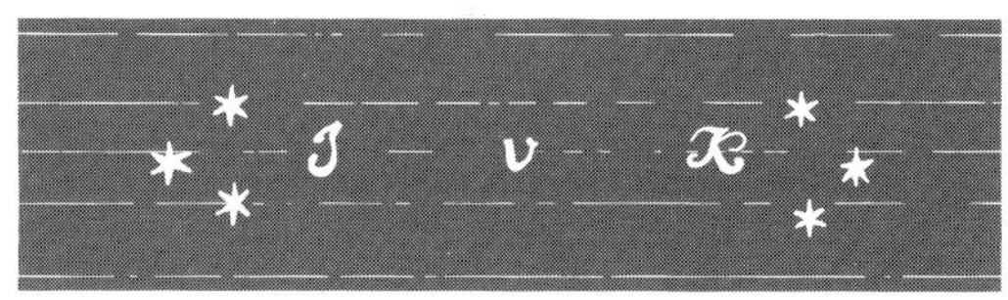

Inscription and decorative stamps on the No.80.

81 Johannes van Keulen (Amsterdam) 1779
742 mm; 19 x 19 mm; ebony

Instrument no. 2 [?]
Inscription side 1
* *
* J * v * K *
* *
Inscription side 2
* *
* 1779 *
* *
Graduation sides
1 *full* 90° - 38°; *range* 80° - 38°; ≥ 10'
2 *full* 90° - 27°; *range* 60° - 27°; ≥ 10'
3 *full* 90° - 13°; *range* 40° - 13°; ≥ 10'
4 *full* 90° - 7°; *range* 20° - 7°; ≥ 10'
Vanes sides
1 506 mm C
2 338 mm C
3 170 mm C
4 84 mm C

Ratio of vane lengths & staff 1 : 2 : 3 : 6 : 9
Condition Fair
Provenance Acquired by the Museum in 1953 at the S.J. van der Vegt Auction in Gorredijk, The Netherlands
Present location Fries Museum, Leeuwarden, The Netherlands, inv.FM. 1953-47
Literature Zijlstra, 'Jakobsstaven', p.4
Remarks No vanes. Graduated for altitude and zenith distance. The end of the staff is pyramid-shaped.

82 Johannes van Keulen (Amsterdam) 1780
752 mm; 20 x 20 mm; ebony

Instrument no. 19 or 61
Inscription side 1
⁎ ⁎ ⁎
⁎ J ⁎ v ⁎ K ⁎ Jan ⁎ Janse ⁎
⁎ ⁎ ⁎
Inscription side 2
⁎ ⁎
⁎ 1780 ⁎
⁎ ⁎
Graduation sides
1 *full* 90° - 40°; *range* 80° - 40°; ≥ 10'
2 *full* 90° - 26°; *range* 70° - 26°; ≥ 10'
3 *full* 90° - 14°; *range* 45° - 14°; ≥ 10'
4 *full* 90° - 7°; *range* 20° - 7°; ≥ 10'
Vanes sides
1 508 mm C
2 338 mm C
3 170 mm C
4 84 mm C
Ratio of vane lengths & staff 1 : 2 : 4 : 6 : 9
Condition Fair
Provenance S. van der Werff, Schiermonnikoog, The Netherlands
Present location Noordelijk Scheepvaartmuseum, Groningen, The Netherlands, inv.5.217-1
Literature
Remarks With modern vanes. Graduated for altitude and zenith distance. The end of the staff is pyramid-shaped. Jan Janse is probably the name of an owner.

83 Not known (Dutch ?) *c.* 1780
NA; 19 x 19 mm; NA; *Vanes* pearwood

Instrument no. 16 [in all four]
Inscriptions None
Vanes sides
1 524 mm
2 377 mm
3 151 mm
4 76 mm
Ratio of vane lengths 1 : 2 : 5 : 7
Condition Poor
Provenance Genootschap Felix Meritis, Amsterdam, The Netherlands
Present location Nederlands Scheepvaartmuseum, Amsterdam, inv.B.25(4)/H.21, since 1984 on loan to the Amsterdams Historisch Museum, Amsterdam, The Netherlands
Literature Brewington, 'Notes on the cross-staff', p.188; Ms NSA 'Lijst van instrumenten', 1971
Remarks Four vanes (width 41 mm) only, with modern staff. Vane four is fitted with a bone horizon.

84 Lambertus Dankbaar (Amsterdam) 1783
750 mm; 18 x 18 mm; ebony

Instrument no. 3
Inscription
1783
⁎ DANKBAAR ⁎
Graduation sides
1 *full* 90° - 38°; *range* 60° - 38°; ≥ 10'
2 *full* 90° - 28°; *range* 60° - 28°; ≥ 10'
3 *full* 90° - 13°; *range* 40° - 13°; ≥ 10'
4 *full* 90° - 7°; *range* 20° - 7 ; ≥ 10'
Vanes sides
1 508 mm C
2 356 mm C
3 166 mm C
4 78 mm C
Ratio of vane lengths & staff 1 : 2 : 4 : 6 : 9
Condition Fair
Provenance
Present location Dr.-Carl-Haeberlin-Friesen-Museum, Wyk auf Föhr, Germany

Literature *Seefahrt einst und heute*, p.79
Remarks With three modern vanes (507, 338 and 168 mm). Graduated for altitude and zenith distance. The end of the staff is pyramid-shaped.

85 Hendrik Mooy (Amsterdam) 1787

861 mm; 16 x 16 mm; ebony; *Vanes* pearwood

Instrument no. 1 [in all parts]
Inscription side 1
1787
⁎ H ⁎ MOOY ⁎
Graduation sides
1 *full* 90° - 39°; *range* 90° - 39°; ≥ 10'
2 *full* 90° - 26°; *range* 60° - 26°; ≥ 10'
3 *full* 90° - 12°; *range* 40° - 12°; ≥ 10'
4 *full* 90° - 7°; *range* 20° - 7°; ≥ 10'
Vanes sides
1 604 mm C
2 396 mm C
3 168 mm C
4 86 mm C
Ratio of vane lengths & staff 1 : 2 : 4½ : 7 : 10
Condition Good
Provenance Found in the attic of the town hall of Sölvesborg, in 1940
Present location Sölvesborgs Museum, Sölvesborg, Sweden, inv.Sb.M.677
Literature
Remarks With original vanes (603, 397, 169, and 88 mm, width 40 mm). Only a fragment of the bone horizon, mounted on vane four, is left. The clamping screw of vane two is missing. Each degree on the zenith distance scale of side three is marked with digits, 75°, 76°, 77° and 78°. Graduated for altitude and zenith distance. The end of the staff is pyramid-shaped. The vanes appear to have been hardly used.

86 Gerard Hulst van Keulen (Amsterdam) 1788

805 mm; 18 x 18 mm; ebony; *Vanes* pearwood

Instrument no. 4 [in all parts]
Inscription side 1
⁎ ⁎
⁎ G.H.V.K. ⁎
⁎ AMST: ⁎
Inscription side 2
⁎ ⁎
⁎ 1788 ⁎
⁎ ⁎
Graduation sides
1 *full* 90° - 38°; *range* 90° - 38°; ≥ 10'
2 *full* 90° - 28°; *range* 40° - 28°; ≥ 5'
3 *full* 90° - 12°; *range* 30° - 12°; ≥ 5'
4 *full* 90° - 6°; *range* 20° - 6°; ≥ 5'
Vanes sides
1 526 mm
2 375 mm
3 151 mm
4 75 mm
Ratio of vane lengths & staff 1 : 2 : 5 : 7 : 11
Condition Good
Provenance Sterrewacht, Leiden, The Netherlands
Present location Museum Boerhaave, Leiden, The Netherlands, inv.A.27.065
Literature *Catalogus 1900*, p.210, no.2008; *Catalogus 1902*, p.286, no.2914; Rooseboom, *Bijdrage*, p.82; Jaspers, 'Over de constructie', p.499n
Remarks With original vanes. Graduated for altitude and zenith distance. The end of the staff is pyramid-shaped. The brass horizon, fitted on vane four, is of later date. The vanes are slightly damaged by woodworm. In 1900 and 1902 the instrument was wrongly dated 1738.

87 Gerard Hulst van Keulen (Amsterdam) 1788

754 mm; 18 x 18 mm; ebony

Instrument no. 5
Inscription side 1
*   *
* G.H.V.K. *
* AMST: *
Inscription side 2
*   *
* 1788 *
*   *
Graduation sides
1 *full* 90° - 38°; *range* 80° - 38°; ≥ 10'
2 *full* 90° - 26°; *range* 60° - 26°; ≥ 10'
3 *full* 90° - 13°; *range* 40° - 13°; ≥ 10'
4 *full* 90° - 7°; *range* 20° - 7°; ≥ 10'
Vanes sides
1 506 mm C
2 338 mm C
3 168 mm C
4 84 mm C
Ratio of vane lengths & staff 1 : 2 : 3 : 6 : 9
Condition Poor
Provenance Ottema Kingma Stichting, Leeuwarden, The Netherlands
Present location Fries Museum, Leeuwarden, The Netherlands, inv.FM.1956-284
Literature Zijlstra, 'Jakobsstaven', p.4.
Remarks No vanes. Graduated for altitude and zenith distance. A hole has been drilled through the staff at the eye-end for a suspension string. Side four has carved marks at intervals of 23; 44; 85; 173; 345 and 516 mm.

88 Lambertus Dankbaar (Amsterdam) 1789

761 mm; 17 x 17 mm; ebony; *Vanes* pearwood

Instrument no. 2 [in all parts]
Inscription side 1
1789
* DANKBAAR *
Graduation sides
1 *full* 90° - 38°; *range* 80° - 38°; ≥ 10'
2 *full* 90° - 26°; *range* 70° - 26°; ≥ 10'
3 *full* 90° - 13°; *range* 45° - 13°; ≥ 10'
4 *full* 90° - 7°; *range* 20° - 7°; ≥ 10'
Vanes sides
1 509 mm C
2 340 mm C
3 168 mm C
4 85 mm C
Ratio of vane lengths & staff 1 : 2 : 4 : 6 : 9
Condition Good
Provenance In 1846 the instrument belonged to the Musée de l'Industrie in Brussels
Present location l'Observatoire Royal de Belgique, Brussels, on loan to the Musées Royaux d'Art et d'Histoire, Brussels, Belgium
Literature Mailly, *Catalogue du Musée*; *Catalogue Exposition*, no.E.2; *Inventaire*, 1, p.50
Remarks With four original vanes. Graduated for altitude and zenith distance. The end of the staff is pyramid-shaped. The bone horizon is missing. An accurate copy of the staff and vanes is in the collection of the Nationaal Scheepvaartmuseum, Antwerp, Belgium, inv.A.S.54.21.1.

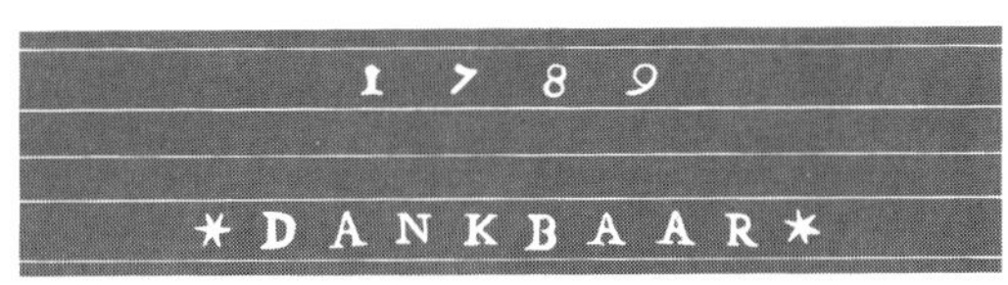

Inscriptions on the No.88.

89 Lambertus Dankbaar (Amsterdam) 1790

757 mm; 18 x 18 mm; ebony

Instrument no. Not found
Inscription side 1
1790
* L * DANKBAAR *
Graduation sides
1 *full* 90° - 38°; ≥ 10'
2 *full* 90° - 26°; ≥ 10'
3 *full* 90° - 13°; ≥ 10'
4 *full* 90° - 7°; ≥ 10'
Vanes sides
1 510 mm
2 340 mm
3 170 mm
4 85 mm
Ratio of vane lengths & staff 1 : 2 : 4 : 6 : 9

Condition Fair
Provenance
Present location Heimatmuseum, Schloss Schönebeck, Bremen-Schönebeck, Germany
Literature
Remarks The modern vanes are marked I, II, III and IV. Graduated for altitude and zenith distance. The end of the staff is pyramid-shaped.

90 Lambertus Dankbaar (Amsterdam) 1790
761 mm; 17 x 17 mm; ebony

Instrument no. 3
Inscription side 1
1790
* L * DANKBAAR *
Graduation sides
1 *full* 90° - 38°; *range* 80° - 38°; ≥ 10'
2 *full* 90° - 26°; *range* 70° - 26°; ≥ 10'
3 *full* 90° - 13°; *range* 40° - 13°; ≥ 10'
4 *full* 90° - 7°; *range* 20° - 7°; ≥ 10'
Vanes sides
1 508 mm C
2 338 mm C
3 168 mm C
4 84 mm C
Ratio of vane lengths & staff 1 : 2 : 4 : 6 : 9
Condition Fair
Provenance Bought by the Museum in 1933 from private owners
Present location Altonaer Museum in Hamburg, Hamburg, Germany, inv.1933-731
Literature Schwindrazheim, 'Der Jakobsstab', no.9
Remarks In 1935 four modern vanes (507, 338, 168 and 83 mm) were made for the instrument, together with a copy of the No.28 box and aperture disk with coloured glass (inv.1934-426). Graduated for altitude and zenith distance.

91 Gerard Hulst van Keulen (Amsterdam) 1791
793 mm; 17 x 17 mm; ebony; *Vanes* pearwood

Instrument no. 2 [in all parts]
Inscription side 1
* *
* G.H.V.K. *
* AMST: *
Inscription side 2
* *
* 1791 *
* *
Graduation sides
1 *full* 90° - 38°; *range* 90° - 38°; ≥ 10'
2 *full* 90° - 28°; *range* 40° - 28°; ≥ 5'
3 *full* 90° - 12°; *range* 40° - 12°; ≥ 5'
4 *full* 90° - 6°; *range* 20° - 6°; ≥ 5'
Vanes sides
1 525 mm
2 375 mm
3 150 mm
4 75 mm
Ratio of vane lengths & staff 1 : 2 : 5 : 7 : $10\frac{1}{2}$
Condition Good
Provenance Koninklijk Instituut voor de Marine, Den Helder, The Netherlands
Present location Nederlands Scheepvaartmuseum, Amsterdam, The Netherlands, inv.S.49(1)/H.16
Literature Brewington, 'Notes on the cross-staff', p.188; Ms NSA 'Lijst van instrumenten', 1971
Remarks With original vanes (width 39 mm) and a new bone horizon made by the Museum in 1919. Graduated for altitude and zenith distance. The end of the staff is pyramid-shaped. Brewington incorrectly dates the staff 1701.

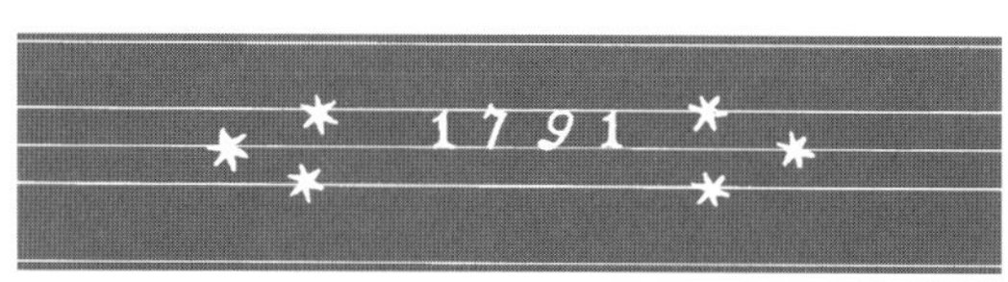

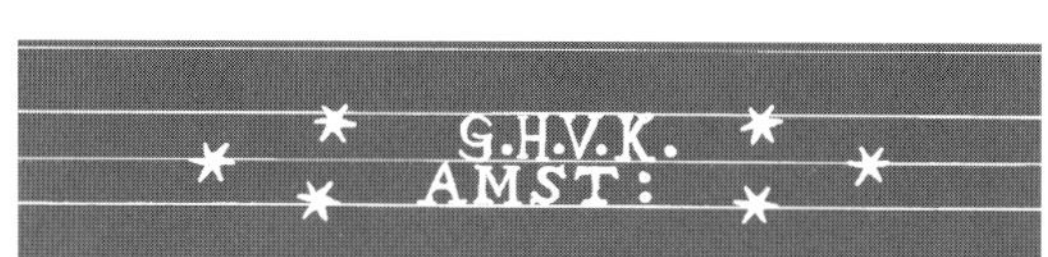

Inscriptions on the No.91.

92 Gerard Hulst van Keulen (Amsterdam) 1793
751 mm; 16 x 16 mm; ebony

Instrument no. Illegible
Inscription side 1
* *
* G.H.V.K. *
* AMST: *
Inscription side 2
* *
* 1793 *
* *
Graduation sides
1 *full* 90° - 38°; *range* 80° - 38°; ≥ 10'
2 *full* 90° - 26°; *range* 60° - 26°; ≥ 10'
3 *full* 90° - 14°; *range* 40° - 14°; ≥ 10'
4 *full* 90° - 7°; *range* 20° - 7°; ≥ 10'
Vanes sides
1 510 mm C
2 342 mm C
3 170 mm C
4 86 mm C
Ratio of vane lengths & staff 1 : 2 : 4 : 6 : 8½
Condition Poor
Provenance
Present location Privately owned, Tiel, The Netherlands
Literature
Remarks No vanes. Graduated for altitude and zenith distance. Side one is badly damaged with three marks carved at intervals of 173 mm.

93 Lambertus Dankbaar (Amsterdam) 1798
NA; NA; ebony

Instrument no. Not found
Inscription side 1
1798
* L * DANKBAAR *
Condition Fair
Provenance
Present location Norsk Sjøfartsmuseum, Oslo, Norway
Literature
Remarks The instrument has not been seen by the author.

94 Hendrik Noordyk (Amsterdam) 1804
NA; NA; NA; *Vanes* NA

Instrument no. Not found
Inscriptions
[H. Noordyk 1804]
Provenance Belonged to Captain G.G. Blankman of Amsterdam (* *c.* 1752 † before 1847), whose son, G. Gerrit Blankman (1785-1859) emigrated to the United States of America between 1813 and 1820
Present location Owned by the Blankman family, U.S.A.
Literature Ms NSA 'The Blankman geneology', 1979
Remarks The instrument has not been seen by the author.

95 Hendrik Noordyk (Amsterdam) 1805
740 mm; 14 x 14 mm; ebony; *Vanes* pearwood

Instrument no. 65 [in all parts]
Inscription side 1
H : NOORDYK
1805
Graduation sides
1 *full* 90° - 39°; *range* 80° - 39°; ≥ 10'
2 *full* 90° - 27°; *range* 60° - 27°; ≥ 10'
3 *full* 90° - 14°; *range* 40° - 14°; ≥ 10'
4 *full* 90° - 7°; *range* 20° - 7°; ≥ 10'
Vanes sides
1 508 mm C
2 338 mm C
3 170 mm C
4 84 mm C
Ratio of vane lengths & staff 1 : 2 : 4 : 6 : 9
Condition good
Provenance School voor Wis- en Zeevaartkunde, Harlingen, The Netherlands
Present location Nederlands Scheepvaartmuseum, Amsterdam, The Netherlands, inv.A.93/H.34
Literature Brewington, 'Notes on the cross-staff', pp.188 & 191; Ms NSA 'Lijst van instrumenten', 1971

Remarks With original vanes (507, 339, 169 and 85 mm, width 37 mm). There is a bone horizon mounted on vane four. Graduated for altitude and zenith distance. The end of the staff is pyramid-shaped.

Inscriptions on the No.95.

Appendix I

Notes on Cross-Staff Makers

+ indicates that the name is entered separately in the list.

F G Initials on a Dutch? cross-staff dated 1714.

H D Initials on a Dutch? cross-staff dated 1682.

I C L Initials on vanes of a Dutch? cross-staff dated *c.* 1740.

J D Z Initials on a cross-staff dated 1765,[1] possibly a misreading of J v K (Johannes van Keulen+).

S O R T Perhaps the monogram of Johan Sort+.

CLAES ANDRIESZ. of Goutum in Friesland, † 1597 on Nova Zembla. The owner and probably also the maker of a cross-staff. He was third mate on one of the Dutch ships, in search of the North-East passage, that sank off Nova Zembla. This resulted in the wintering of the crews, among them Willem Barentsz. and Jacob van Heemskerck, in 1596-1597. The cross-staff was left behind in 1597 and rediscovered in 1871.[2]

EVERT APEGENDAM of Amsterdam. Cross-staff maker, became citizen of Amsterdam in 1730.[3]

WILLEM JANSZOON [BLAEU]. ✳ Alkmaar 1571 † Amsterdam 1638. Apprentice to Tycho Brahe at his observatory Uraniborg on the Danish island of Ven. Settled in Amsterdam in 1598 or 1599 as globemaker and astronomical instrument maker. He was a printer and publisher of charts, maps, atlases and books. In 1607 he made an astronomer's cross-staff for Adriaan Metius that was an improved version the staff of Gemma Frisius. The name Blaeu was added in 1621 to distinguish him from his rival, Jansonius.[4]

P. BOEKENES. Made or owned a cross-staff dated 1707 and signed P. BOEKENES. Can perhaps be identified as the publisher and bookseller Paulus Boekenes, who lived *op de Beurs* in Rotterdam, and who was a member of the Booksellers Guild from 1681 to 1712.[5]

G.D. BOM. Name on a Van Keulen cross-staff dated *c.* 1710, probably that of an owner. Perhaps the G.D. Bom whose father auctioned the belongings of the liquidated firm of van Keulen in 1885 and 1919.[6]

WIGGERT COERTEN of Amsterdam, ✳ *c.* 1632. Cross-staff maker, living on the Brouwersgracht. Married Geertruyt Alders in 1658.[7]

JOHN COGGS I of London, active 1718-1733, in partnership with his former apprentice William Wyeth. Mathematical instrument maker, advertised fore-staffs and other instruments. In 1729 Coggs was established at the Globe and Sun over against St. Dunstan's Church in Fleet Street, London.[8] According to Taylor, John Coggs II advertised fore-staffs in the 1730s. This seems unlikely as Coggs II was not apprenticed to Benjamin Cole until 1759.[9]

FRANCIS COOKE of London, instrument maker, active 1590-1596. From his house in Mark Lane he made and sold the 'staves' which were designed by Thomas Hood.[10]

JACOB JACOBSZ. CROEGER of Amsterdam, ✳ *c.* 1653. Cross-staff maker, living in the Boerensteeg. Married Catrijntje Hendricx in 1673.[11]

THOMAS CROSBY & JOHN ROBINSON of London, active *c.* 1740. Before 1743 they were established at the Lower End of Fair-Street upon Horse-ly-down in Southwark, London. They were sellers of mathematical and optical instruments, and teachers. Advertised fore-staffs.[12]

LAMBERTUS DANKBAAR of Amsterdam, ✳ *c.* 1733 † before 1806[13]. Cross-staff maker living on the Oosterse Kaaij. Married, in 1768, Anna Maria Simmers[14]. Became citizen of Amsterdam in 1769.[15] Signed ★ L ★ DANKBAAR ★ or merely ★ DANKBAAR ★.

ANTONIJ DAVIDS of Amsterdam, ✳ *c.* 1657. Cross-staff maker living in the Runstraat. Married Catrijntje Jacobz. in 1680.[16]

MATIJS DOMINICUS of Amsterdam. Cross-staff maker; became citizen of Amsterdam in 1753.[17]

HENDRIK DONCKER of Amsterdam, ✳ *c.* 1626 † Amsterdam 1699. Signed ★H ★ DONCKER★. In 1647 he was admitted to the Booksellers Guild. Married Elisabeth Abrahams Rensing in 1648.[18] In 1651 Doncker established his firm in the Nieuwe Brugsteeg *In 'T Stuurmans gereetschap*, living there until his death. In 1655 he advertised as bookseller and cross-staff maker.[19] In 1664 Doncker advertised as selling 'all sorts of cross-staffs; mirror-staffs; cross-bows; English and Dutch quadrants; nocturnals, plane-scales by several masters ... etc.'.[20] Doncker sold his house to Johannes van Keulen+ in 1693, who then probably let it to him. Doncker's son continued the business for some years.

HENDRIK II DONCKER of Amsterdam, ✳ *c.* 1664 † *c.* 1739, son of Hendrik Doncker+. Continued his father's business on the same premises in 1699. The house and stock were auctioned in 1720.[21] It cannot be confirmed that, like his father, he made cross-staffs.

DIRCK EVERTSZ. from Londen in Ditmarssum. Cross-staff maker, living in the Minnebroedersteeg. Widower of Jannetje Jans, and married in 1654, Clara Lourus.[22]

PIETER GOOS of Amsterdam, ✳ 1614/16 † 1675. Chartmaker and seller. In 1659 he advertised as selling newly invented cross-staffs divided in degrees and minutes.[23] After his death his business went to Jacobus Robyn+.

HENDRIK GOOSSENS of Amsterdam, ✳ *c.* 1649. Cross-staff maker, living in the Nieuwe Zijds Kapelsteeg. Married in 1677 Sybrigh Jans.[24]

ISAAC DE GRAAF of Amsterdam, ✳ *c.* 1668 † Amsterdam 1743. Son of Abraham de Graaf, examiner of the mates who sailed for the Amsterdam Chamber of the VOC. When Isaac married Sanderina de Brauw in 1708, he was living on the Brouwersgracht.[25] He was appointed as draughtsman of VOC charts in September 1691. De Graaf became official chartmaker to the VOC in 1705, after Joan II Blaeu's resignation. He held this position until his death in 1743 and was succeeded by Johannes van Keulen+.[26] In 1730 and 1731 he sold an unknown number of cross-staffs to the VOC. However, in December 1731, 30 cross-staffs were returned to De Graaf by the Company.[27]

HARMEN GROOTENHOFF from Wessum, ✳ *c.* 1664. Cross-staff maker, living on the Zeedijk in Amsterdam. Married Grietje Bruyningh in 1690.[28]

ANTONIJ GUNTER from Oldenburg, ✳ *c.* 1653. Cross-staff maker, living in Oude Zijds Kapelsteeg in Amsterdam. Married Gerritje Willems in 1678.[29]

HAMON of St Malo. Made or owned a cross-staff dated 1687, and signed HAMON A ST MALO.

JOSEPH HARRIS of London, active 1724-1764. Mathematical teacher and instrument maker. In 1729 he was established at Two Civet Cats, Crane Court, Fleet Street, London. In 1730, in his *A Treatise of Navigation*, Harris included a chapter on the description and use of a new fore-staff which he had designed, and which was sold by Thomas Heath+. In 1748 Harris was Master of the Mint at the Royal Mint, Tower Hill, London.[30]

GERRIT HASEBROEK of Amsterdam. Son of Jochem Hasebroek+, who resided at the same address.[31] Signed *G ★ Hasebroek*. Became citizen of Amsterdam in 1746.[32] Listed in 1767 as cross-staff maker in the Nieuwe Brugsteeg.

JOCHEM HASEBROEK of Amsterdam ✳ 1682 † Amsterdam 1756, buried in the New Church on 24 February. Married Hendrikje Doncker (✳ Weesp 1683) in 1708.[33] Signed *J ★ Hasebroek*. In 1708 Hasebroek was an art dealer, and in the same year was admitted to the Booksellers Guild. Hasebroek was a bookseller and cross-staff maker in the Nieuwe Brugsteeg *In 't Stuurmans Gereedschap*, possibly the business left by Jacobus Robyn+. On the cartouche of a chart dated 1716 the business was named *In de Stuurman*. Hasebroek lived on the premises. In 1750 he bought the copper plates of the late Isaac Swigters, who had bought them from Johannes Loots+. Loots had, around 1700, acquired the plates from Jacobus Robyn+.[34]

THOMAS HEATH of London, † London 1773. Active 1720-1751 as mathematical and optical instrument maker. Apprenticed to Benjamin Scott in 1712, made Freeman of the London guild, the Grocers' Company, in 1719-1720. From 1720 to 1747 he was established Near the Fountain Tavern in the Strand, and 1734-1746 at the Hercules & Globe near Beaufort Buildings in the Strand, probably the same premises. In 1730 Heath made and sold a new fore-staff designed by Joseph Harris+. The latter's *A Treatise of Navigation*, containing a chapter on the description and use of his invention, was published by Heath that year. In 1750 Heath was established at the Hercules & Globe near Exeter Exchange in the Strand. In 1758 he went into partnership with his son-in-law and former apprentice, Tycho Wing.[35]

WILLIAM HINTON of New York. In 1772 he is recorded as making and selling all sorts of mathematical instruments and cross-staffs.[36]

GERARD HULST VAN KEULEN, ✳ Amsterdam 1733 † Amsterdam 1801. Signed G.H.V.K. AMST:. Son of Johannes II van Keulen+. Gerard Hulst was admitted to the Booksellers Guild on 15 December 1755. From 1756 to 1779 he and his brother Cornelis Buys van Keulen ran the firm of Johannes van Keulen en Zoonen, a combination of an anchor foundry and the original publishing, printing, instrument and chart making firm. After the death of Cornelis Buys in 1779, the firm was split. The anchor foundry was continued by the widow of Cornelis Buys, while Gerard Hulst associated the other section of the firm with Christiaan Hulst & Comp, and continued the business under his own name until his death.[37]

WEDUWE G. HULST VAN KEULEN of Amsterdam, active 1801-1885. After Gerard Hulst died in 1801 his widow, Anna Hendrina Calkoen, continued the firm until her death in 1810. Her grandson, Johannes Gerard Hulst van Keulen van de Velde, became the next owner, selling the firm out of the family in 1823. It was run by outsiders under the above name until liquidated in 1885. Until 1818 the firm advertised as cross-staff makers;[38] however, no examples are known to exist. In 1853 the firm participated in a maritime exhibition in Amsterdam.[39] One of the items displayed was 'an old cross-staff'. However, when the stock of the firm was auctioned in 1885 and 1919, no cross-staff was among the numerous instruments.[40]

LEENDERT HUSTING of Amsterdam, † before 1785.[41] Cross-staff maker, became citizen of Amsterdam in 1742.[42]

LOURENS JACOBSZ. of Amsterdam, ✳ *c.* 1645. Cross-staff maker, living on the Fluwelenburgwal. In 1666, married Engeltje Reynierse.[43] In 1667 the estate of Claes Hendricksz. Gietermaker owed Lourens Jacobsz. *f* 12 for four cross-staffs.[44]

GERRIT JAGER of Amsterdam, ✳ *c.* 1744. Cross-staff maker, living in the Lindenstraat. Married in 1771 Jacoba Oostens Blankenharts.[45] Became citizen of Amsterdam in 1772.[46]

JAN JANSE Name on a Van Keulen cross-staff dated 1780, probably that of an owner.

ANDRIES JANSEN of Amsterdam. In 1600 delivered cross-staffs to the ships of Van Neck for the second voyage of the Dutch to the East Indies.[47]

CLAAS JANSZ. † before 1750. Cross-staff maker, became citizen of Amsterdam in 1723.[48]

JAN JANSZ. of Amsterdam, ✳ *c.* 1656. Cross-staff maker, living on the Baangracht. Married Amerencia Jans in 1681.[49]

JACOB JANSZ. KAASKAMER of Amsterdam, ✳ *c.* 1632. Cross-staff maker, living on the Stroomarket. In 1659 married Catharina Uylenburghs.[50] Perhaps the same as cross-staff maker Jacob Jansz., as mentioned in 1674.[51]

I.I. KAES H [or B] [illegible]. Inscription on an American? cross-staff dated 1666.

GERARD VAN KEULEN ✳ Amsterdam 1678 † Amsterdam 1726. Signed *G v K*. Son of Johannes I van Keulen+ and father of Johannes II van Keulen+. Gerard was admitted to the Booksellers Guild on 24 February 1704, the same year he took over from his father the business which he continued until his death.[52]

JOHANNES I VAN KEULEN ✳ Deventer(?) *c.* 1654 † Amsterdam 1715. Signed I ★ V ★ K, ★ *J v K* ★ and *J* ★ *v* ★ *K*. Father of Gerard van Keulen+. On 28 August 1678 Johannes, who was not a native of Amsterdam, was admitted to the Booksellers Guild. It is not unlikely that he produced charts before that year, possibly as early as 1674.[53] In 1678 he settled as bookseller and cross-staff maker in the Nieuwe Brugsteeg naming his firm *In de Gekroonde Lootsman*. In 1679 he was briefly associated with Jacobus Robyn+. In 1680 he advertised in the French edition of his *Sea-Atlas* as *Libraire & faiseur de roye astronomique*.[54] In 1693 Johannes acquired the house and business of Hendrik Doncker+. Johannes employed two men who were instrumental in making the Van Keulen firm flourish, namely Claes Jansz. Vooght († 1696) and Johannes van Loon († 1686). Vooght was mathematician, surveyor and teacher in the art of navigation while Van Loon, mathematician, astronomer and geometer, called himself engraver and sea chart maker. Both were authors of several early and important books published by Van Keulen. Vooght invented the *Quandrans Astronomicus* which was drawn by Van Loon, and of which a few copies survive.[55]

JOHANNES II VAN KEULEN of Amsterdam, † Amsterdam 1755. Signed I ★ V ★ K and *J* ★ *v* ★ *K*. Son of Gerard van Keulen+ and father of Gerard Hulst+. In 1724, one year before the death of his father, Johannes II became joint owner of the Van Keulen firm. A year later, on 26 November 1725, he was admitted to the Booksellers Guild. As cartographer he appears less important than his father and grandfather; as instrument maker he very likely surpassed them both in production. In 1743 Johannes II was appointed official cartographer of the VOC,[56] the position which was inherited by his son. Prior to that appointment, starting in 1731, he sold cross-staffs to the VOC (see Appendix III). Johannes II was, after Jan van Musschenbroek and Benjamin Ayres, the third Dutch instrument maker to provide octants to the Company, from 1751 onwards.[57] After his death the firm was continued by his two sons, Cornelis Buys and Gerard Hulst+.[58]

MIGHIEL KONING of Amsterdam, ✳ *c.* 1711 † before 1780. Became citizen of Amsterdam in 1753.[59] When he married Anna Bastert in 1753, he was living on the Haarlemmerdijk.[60]

ANTHONY LAMB of New York, ✳ London 1703 † New York 1784. Mathematical instrument maker At the sign of the Compass and Quadrant in New York. He was deported to the American colonies in 1724 and in 1730 advertised as maker and seller of all sorts of instruments, including fore-staffs.[61]

JOHANNES LOOTS ✳ Amsterdam 1665 or 1666 † Amsterdam 1726. Signed *★J ★ Loots★*. Bookseller and cross-staff maker. He was an apprentice of Hendrik Doncker+. When he married Sara Swigters in 1691, he was a mathematical instrument maker living on the Zeedijk.[62] Around 1693 the couple moved to the Nieuwe Brugsteeg to *In de Jonge Lootsman*, the fifth house from the Nieuwe Brug. On 5 May 1693 Loots was admitted to the Booksellers Guild and on 1 April 1695 he started selling charts. Although his earliest known cross-staff is dated 1696, the first mention of him as a maker is from 1697.[63] In 1721 Loots acquired a house on the Binnen Amstel. After his death in 1726, his widow Sara Swigters continued the firm until she died in 1743. Sara's brother, Isaac Swigters, in turn continued the firm until he died in 1750, after which it was split up. The inventory of the merchandise left by Swigters was drawn up by notary public Arnoldus de Ridder on 30 September 1750, and included 14 cross-staffs, of which two were made of ebony.[64] An advertisement on 30 March 1751 mentions the sale of part of the firm's property, in which was ebony and pearwood, possibly for making cross-staffs.[65] Hendrik Mooy+ (1715-1765), cousin twice removed of Sara Swigters, bought the instruments and continued the instrument section of the firm *In de Jonge Lootsman*, on the premises where Loots had started around 1693.

JACOB THEUNIS LOOTSMAN of Amsterdam, † 1679. Son of chartmaker Anthony Jacobsz. Lootsman. Jacob Theunis, who published charts and sea-atlases in Amsterdam between 1666 and 1676, advertised as bookseller and cross-staff maker *In de Lootsman*. After 1677, the publishing firm became Jacob & Casparus Lootsman and no longer advertised as making cross-staffs.[66]

JOHANNES LUBS of Amsterdam. Cross-staff maker. When he became citizen of Amsterdam, in 1746, he was already married to Barbara Martens.[67]

ABRAHAM MANUEL of Amsterdam, ✳ *c.* 1718 † before 1761. Became citizen of Amsterdam in 1747,[68] when he was living *Op 't Water*. Married Grietje Romeijn in 1747. She, as a widow, remarried in 1761.[69]

HENDRIK MOOY, ✳ Amsterdam 1715 (Christened 18 October) † Amsterdam 1765 (buried 8 November). Signed ★H ★ MOOY★. On 11 December 1750 Mooy became citizen of Amsterdam[70] and on 14 December 1750 he was admitted to the Booksellers Guild. The same year he established his business *In de Jonge Lootsman*, the firm which was founded by Johannes Loots+ around 1693. It appears that he did not live on the premises. When, on 25 April 1751, Mooy married Alida Hamburg (1721-1795) he was living in the Reguliersbreestraat. His widow, who in 1769 married an old friend of her husband Carel ten Ham, continued the firm under the name *Erven van Hendrik Mooy*, probably until her death.[71] Nevertheless Hendrik's name survived for many years. In a 1767 list of shop owners, the firm is still entered as *Hendrik Mooy graadboog maker in de Nieuwe Brugsteeg*[72] and the two known cross-staffs signed Mooy are dated after his death.

HENDRIK NOORDYK of Amsterdam, ✳ 1753 † 1828. Mathematical instrument maker and cross-staff maker. Signed H: NOORDYK. Lived in the Kalverstraat in 1778, when he married Anna Maria Kneus. Became citizen of Amsterdam in 1782[73]. A widower, he married

Catharina Roode in 1790. He then lived in the Keijzerstraat near the Boomsloot.[74] Besides two cross-staffs, a mathematical ruler by Noordyk is known.[75]

JAMES PATERSON of Edinburgh. Active 1679-1693 as mathematician and mathematical instrument maker. From 1681-1686 he was established at the Sign of the Sea Cross-staff and Quadrant, Cowgate, and in 1689 at the Sign of the Sea Cross-staff and Quadrant, Reid House, middle of Leith Wynd. In 1687 he advertised cross-staffs, quadrants, scales, sea compasses, etc. Paterson was succeeded by his nephew John Mann.[76]

HENDRIK PIETERSZ. from Elburg, † before 1707.[77] Signed ★ H . P ★. Cross-staff maker, living in the Haarlemmerstraat in Amsterdam. Widower of Jannetje Jans; in 1670 married Aaltje Jacobs in Amsterdam.[78]

JACOBUS ROBYN of Amsterdam, ✳ *c.* 1649 † - between 1707 and 1717. Robyn married Jannetje Tol in Amsterdam on 15 December 1673. On 26 January 1674 he was admitted to the Booksellers Guild as book and chart seller in the Nieuwe Brugsteeg, at *De Stuurman*. In 1679 he was briefly associated with Johannes I van Keulen+. According to an advertisement in his *Zee-Spiegel* of 1688, he then made and sold cross-staffs. His firm was probably continued by Joachim Hasebroek+.[79]

LUCAS VAN SCHAGEN of Amsterdam. Cross-staff maker, became citizen of Amsterdam in 1732.[80]

JACOBUS SCHRAVELDER of Amsterdam. Cross-staff maker, son of Anthony, became citizen of Amsterdam in 1747.[81]

JOHN SELLER [SELLARS] of London, † 1697. Active 1660-1697 as mathematical instrument maker, map maker, compass maker, globe and book seller. Advertised cross-staffs for sale. In 1660 he was established at the sign of the Mariners Compass and Hour Glass at the Hermitage Stairs in Wapping, and 1669-1696 at the Mariners Compass at the Hermitage Stairs in Wapping. Seller was succeeded by his widow Elizabeth, and his son Jeremiah.[82]

JOHAN SORT from Garresyn, ✳ *c.* 1641. Cross-staff maker and mathematical instrument maker, living in the Oude Zijds Kapelsteeg in Amsterdam. Married in 1670 Anna Reyniers from Essen.[83] He was living on the Zeedijk when, in 1688, as a widower, he married Katrina Janse Grootenhoff.[84] Perhaps the maker of two cross-staffs, dated 1680 and 1684, which have the monogram S O R T+ applied with a single, specially-made stamp.

ABRAHAM TADE of Amsterdam, ✳ *c.* 1677. Cross-staff maker, living in the Haantjeshoeksteeg. Signed *A ★ Tade*. Married Jannetje Muijs in 1706.[85]

EN S TELMO Name on an American? cross-staff dated 1731, and signed EN S TELMO.

THOMAS TUTTELL [TUTTLE] of London, † 1702. Active 1695-1702 as mathematical instrument maker. Signed *Thomas Tuttell Londini fecit*. From 1700-1702 hydrographer and mathematical instrument maker to William III. In 1697 he was established At the Kings Arms and Globe at Charing Cross. Tuttell made a cross-staff dated c 1700.[86]

GERRIT GERRITSZ. TYHOF from Oldenzeel, ✳ *c.* 1638. Cross-staff maker, living in the Kapelsteeg in Amsterdam, when he married Jannetje Jacobs in 1668.[87]

ALBERT VISS[CH]ER of Amsterdam, † before 1747. Cross-staff maker. When, in 1725, he became citizen of Amsterdam he was already married to Catharina Voort.[88]

HENDRIK VOUWENS of Amsterdam, † before 1765. Cross-staff maker, became citizen of Amsterdam in 1735.[89]

ANDREW WAKELY [WAKERLY] of London. Active 1633-1669 as mathematician and mathematical instrument maker, at Redriff Wall near Cherry Garden Stairs. He was succeeded by James Atkinson. Wakely advertised that he sold fore-staffs.[90]

JOH. WARMELO Listed in 1742 as cross-staff maker in the Utrechtsestraat in Amsterdam.[91]

PIETER WILLEMSZ. of Hoorn. In 1620 boxmaker Willemsz. provided cross-staffs to the Dutch East India Company.[92]

BENJAMIN WORKMAN of London. Active after 1726 as mathematical instrument maker, Next Tom's Coffee House in Russell-Street, Covent Garden. He advertised fore-staffs, among other instruments.[93]

Appendix II

Values of cross-staffs 1576-1804

1576	13s. 4d.[1]
1595-1650	*f* 2-12[2] [two guilders and twelve *stuivers*, there are twenty stuivers to a guilder]
1655-1675	*f* 2-11 *graetboogh met een kopere visier en houte horisont*[3] [... with a brass aperture disk and wooden horizon vane]
1673	*f* 2-10 *graetboogh met een kopere visier en houte horisont*[4]
1675-1731	*f* 3 *graetboog met koopere visier en houte horisont*[5]
1732	*f* 4-5 paid for cross-staffs made by Isaac de Graaf[6]
1731-1747	*f* 5 *graadboogh met kopere visier en dubbele kopere schroeven*[7] [... with a double set of brass screws]
1747-1791	*f* 5-4 *ordinaire graadboog met duble vizier*[8]
1759	*f* 3-11 *een graadstok*[9]
1753-1780	*f* 5-4 *graadboog met vizier te waarlo*[10] [... with a spare aperture disk]
1765	*f* 1-4 *twee kruise voor een gradestok*[11] [for two vanes]
1777	*f* 5 *graadboog met dubbele bossen* [by *bos* in this case, the brass clamping screw and plate are meant, of which two are provided] *dubbele vizieren* [two aperture disks] *en tekening* [graduation].
	f 3-10 *graadboog met dubbele bossen, vizieren en tekening*
	f 2-10 *graadboog met ordinaire bossen, vizieren en tekening*[12]
1788-1791	*f* 5-4 *graadboog met vizieren ter waarlo*[13] [... with spare aperture disks]
1800-1804	*f* 5-4 *graadboog met vizieren te waarlo*[14]

Appendix III[1]

Purchases of cross-staffs by the VOC between 16 May 1728 and 30 June 1748

16 May 1728 to 15 May 1732[2]

30/06/1730 Isaac de Graaf; ? cross-staffs

31/12/1730 Isaac de Graaf; ? cross-staffs

28/02/1731 Isaac de Graaf; ? cross-staffs

30/06/1731 Isaac de Graaf; ? cross-staffs (for the Rotterdam Chamber of the VOC)

31/08/1731 Joannes van Keulen; 31 cross-staffs

31/12/1731 returned to Isaac de Graaf 30 cross-staffs for *f* 127-10[3] (*f* 4-5 per cross-staff)

15/05/1732 Johannes van Keulen; ? cross-staffs

16 May 1732 to 31 May 1736[4]

15/05/1733 Johannes van Keulen; ? cross-staffs

31/01/1734 Johannes van Keulen; 49 cross-staffs for *f* 5 each

30/11/1734 Johannes van Keulen; 31 *graadbogen met dubbele platen*[5] for *f* 5 each

31/01/1735 Johannes van Keulen; 30 *graadbogen met dubbele platen* for *f* 5 each

30/06/1735 Johannes van Keulen; 29 *graadbogen met dubbele platen* for *f* 5 each

31/12/1735 Johannes van Keulen; 29 *graadbogen met dubbele platen* for *f* 5 each

30/06/1736b Johannes van Keulen; 66 *graadbogen met dubbele platen* for *f* 5 each + 66 *visiers* for 5 *stuivers*[6] each

1 June 1736 to 31 May 1740[7]

28/02/1737 Johannes van Keulen; 88 *graadbogen met dubbele bossen* for *f* 5 each

31/05/1737 Johannes van Keulen; 33 *graadbogen met dubbele platen* for *f* 5 each

31/01/1738 Johannes van Keulen; 42 *graadbogen met dubbele platen* for *f* 5 each
Johannes van Keulen; 36 cross-staffs

15/05/1738 Johannes van Keulen; 32 cross-staffs for *f* 5 each

31/12/1738 Johannes van Keulen; 78 *graadbogen met dubbele platen* for *f* 5 each

15/05/1739 Johannes van Keulen; 15 *graadbogen met dubbele platen* for *f* 5 each

31/01/1740 Johannes van Keulen; 70 *graadbogen met dubbele platen* for *f* 5 each

1 June 1740 to 15 May 1744[8]

30/06/1740 Johannes van Keulen; 15 *graadbogen met dubbele platen* for *f* 5 each

30/11/1740 Johannes van Keulen; 34 *graadbogen met dubbele platen* for *f* 5 each

31/01/1741 Johannes van Keulen; 22 *graadbogen met dubbele platen* for *f* 5 each

15/05/1741 Johannes van Keulen; 13 *graadbogen met dubbele platen* for *f* 5 each

31/10/1741 Johannes van Keulen; 28 *graadbogen met dubbele platen* for *f* 5 each

31/12/1741 Johannes van Keulen; 29 *graadbogen met dubbele platen* for *f* 5 each

31/05/1742 Johannes van Keulen; 9 *graadbogen met dubbele platen* for *f* 5 each

31/12/1742 Johannes van Keulen; 42 *graadbogen met dubbele platen* for *f* 5 each

31/01/1743 Johannes van Keulen; 21 *graadbogen met dubbele platen* for *f* 5 each

15/05/1743 Johannes van Keulen; 15 *graadbogen met dubbele platen* for *f* 5 each and *vizieren* for 5 *stuivers* each

30/11/1743 Johannes van Keulen; 31 cross-staffs for *f* 5 each

29/02/1744 Johannes van Keulen; 9 cross-staffs for *f* 5 each

16 May 1744 to 31 May 1748[9]

31/05/1744 Johannes van Keulen; 17 *graadbogen met dubbele platen* for *f* 5 each

30/11/1744 Johannes van Keulen; 39 *graadbogen met kopere vizieren en dubbele platen* for *f* 5 each

30/11/1745 Johannes van Keulen; 12 *graadbogen met dubbele platen* for *f* 5 each

15/05/1746 Johannes van Keulen; 25 *graadbogen met dubbele platen* for *f* 5 each

Johannes van Keulen; 30 *graadbogen met dubbele platen* for *f* 5 each

31/07/1746 Johannes van Keulen; 30 *graadbogen met dubbele platen* for *f* 5 each

30/11/1746 Johannes van Keulen; 12 *graadbogen met dubbele platen* for *f* 5 each

28/02/1747 Johannes van Keulen; 24 *graadbogen met dubbele platen* for *f* 5 each

30/04/1747 Johannes van Keulen; 30 *graadbogen met dubbele platen* for *f* 5 each

30 June 1748[10]

30/06/1748 Johannes van Keulen; 2 cross-staffs for *f* 5 each

Appendix IV

Instrument numbers found on surviving cross-staffs, in the sequence of the Checklist. As the majority of the staffs were made by the firm of Van Keulen, these are listed separately, giving an idea of the use of numbers by one firm.

No.	Date	Maker	All	Van Keulen
6	1677	Pietersz.	7	
11	*c.*1690	not known	8	
12	1691	Van Keulen	H	H
13	1696	Loots	I	
14	1698	Doncker	4	
15	1699	Tade	H	
18	*c.*1700	not known	4	
19	1701	Van Keulen	H, E or 7	H, E or 7
23	1706	Van Keulen	2	2
24	1707	Boekenes	4 H	
29	1714	Loots	10	
32	1716	Loots	11	
33	1717	Van Keulen	K	K
35	1720	Hasebroek, J.	H3	
38	1724	Van Keulen	5	5
41	1731	not known	3	
44	1734	Hasebroek, J.	HV [or A]	
48	*c.*1740	not known	3	
49	1740	Van Keulen	18 or 81	18 or 81
53	*c.*1743	Hasebroek, J.?	H	
54	1745	Van Keulen	7	7
55	1748	not known	10	
56	1749	Van Keulen	6 or 9	6 or 9
57	1749	Van Keulen	2 DO	2 DO
59	1750	not known	5	
60	1751	Van Keulen	2	2
62	1753	Hasebroek, J.	1	
63	1755	Van Keulen	2	2
64	1756	Van Keulen	O	O
66	1758	Van Keulen	1	1
67	1760	Van Keulen	12	12
70	1765	Hasebroek, G.	ʏ	
72	1765	Van Keulen	10	10
74	1775	Van Keulen	1.0	1.0
75	1775	Mooy	2	
76	*c.*1775	not known	4	
77	1776	Van Keulen	15	15
78	1777	Van Keulen	4	4
79	1777	Van Keulen	8	8
80	1778	Van Keulen	5	5
80	1778	Van Keulen	16	16
81	1779	Van Keulen	2 [?]	2 [?]
82	1780	Van Keulen	19 or 61	19 or 61
83	*c.*1780	not known	16	
84	1783	Dankbaar	3	
85	1787	Mooy	1	
86	1788	Van Keulen	4	4
87	1788	Van Keulen	5	5
88	1789	Dankbaar	2	
89	1790	Dankbaar	3	
91	1791	Van Keulen	2	2
95	1805	Noordyk	65	

Appendix V

The lengths of surviving cross-staffs, in mm, in the sequence of the Checklist. Fragments, and staffs which have obviously been shortened considerably, have not been included. A + indicates that shortening is uncertain or very little. The Dutch and the Van Keulen staffs have been listed separately. Comparison shows that the development of the average length of these corresponds with the other staffs.

No.	Year	All	Dutch	Van Keulen
1	≤1596	700+	700+	
2	≤1622	867+		
3	1661	1010		
4	1666	915		
6	1677	1005	1005	
7	1680	960+	960+	
8	1682	1007+		
9	1684	915+	915+	
10	1687	915		
11	c.1690	830+		
12	1691	582	582	582
13	1696	770+	770+	
14	1698	870	870	
15	1699	826	826	
16	1700	660+	660+	600+
17	c.1700	796		
19	1701	872	872	872
20	1702	755+	755+	755+
21	1703	754	754	
22	≤1703	837+		
23	1706	732	732	732
24	1707	751		
26	1710	713+	713+	713+
27	c.1710	738	738	738
29	1714	734	734	
30	1714	827+		
32	1716	828	828	
33	1717	744	744	744
34	≤1717	705		
35	1720	658	658	
36	c.1720	722+	722+	
37	1721	764+	764+	764+
38	1724	838	838	838
39	c.1727	740+		
40	1729	720	720	
41	1731	671		
42	1732	914		
43	1733	731+	731+	731+
44	1734	830	830	
45	c.1735	830		
47	1736	745	745	745
49	1740	858	858	858
50	1741	719	719	
51	1743	837	837	
54	1745	744	744	744
55	1748	835		
56	1749	762	762	762
57	1749	742	742	742
59	1750	854		
60	1751	750+	750+	750+
61	1752	752+	752+	
62	1753	843	843	

63	1755	750	750	750
64	1756	743	743	743
65	1758	657	657	
66	1758	749	749	749
67	1760	753	753	753
68	1762	743	743	743
69	1764	681	681	681
70	1765	718	718	
71	1765	737+	737+	737+
72	1765	658	658	658
73	1765	667		
74	1775	747	747	747
75	1775	755	755	
76	c.1775	619		
77	1776	670	670	670
78	1777	755	755	755
80	1778	774	744	744
81	1779	742	742	742
82	1780	752	752	752
84	1783	750	750	
85	1787	861	861	
86	1788	805	805	805
87	1788	754	754	754
88	1789	761	761	
89	1790	757	757	
90	1790	761	761	
91	1791	793	793	793
92	1793	751	751	751
95	1805	740	740	

Lengths given in some contemporary navigational manuals are listed below. Comparison shows that staffs were slightly longer in theory than in practice.[1]

6 palms	Cortes 1551	1290	mm
3-3½ feet	Coignet 1580	849-990	mm
4 palms	Zamorano 1581/98	860	mm
3-4 feet	Blundeville 1594	914-1219	mm
5 palms	Garcia de C. 1606	1075	mm
3-3½ feet	De Decker 1631	849-990	mm
3-4 pieds	Fournier 1643	972-1296	mm
4 feet	Colom 1652	1132	mm
3-3½ feet	De Graaf 1658	849-990	mm
2-2½ feet	Vooght 1684	566-707	mm
2½-3 pieds	Bouguer 1698	810-972	mm
2, 3½-4½ feet	De Vries 1702	566, 990-1273	mm
3 feet	Harris 1730	914	mm
3-3½ feet	Gietermaker 1742	849-990	mm
3 feet	Robertson 1754	914	mm

Appendix VI

The widths of the surviving staffs, in mm, in the sequence of the Checklist. The Dutch and the Van Keulen staffs have been listed separately. Their widths appear to follow the same development as the others.

No	Year	All	Dutch	Van Keulen
1	≤1596		13	
2	≤1622	15		
3	1661	14½		
4	1666	12		
5	≤1676	13		
6	1677	14	14	
7	1680	12	12	
8	1682	14		
9	1684	13	13	
10	1687	13		
11	c.1690	14		
12	1691	13	13	13
13	1696	14	14	
14	1698	14	14	
15	1699	13	13	
16	1700	14	14	14
17	c.1700	14		
18	c.1700	18		
19	1701	14	14	14
20	1702	15	15	15
21	1703	14	14	
22	≤1703	16		
23	1706	17½	17½	17½
24	1707	15	15	
26	1710	14½	14½	14½
27	c.1710	14	14	14
29	1714	16	16	
30	1714	19		
32	1716	16	16	
33	1717	15	15	15
34	≤1717	17		
35	1720	18	18	
36	c.1720	17	17	
37	1721	15	15	15
38	1724	16	16	16
39	c.1727	18		
41	1731	16		
43	1733	16	16	16
44	1734	17	17	
45	c.1735	19		
46	≤1735	17	17	
47	1736	15	15	15
48	c.1740	15		
49	1740	15	15	15
50	1741	16	16	
51	1743	17	17	
52	≤1743	15		
53	≤1743	14	14	
54	1745	16	16	16
55	1748	15		
56	1749	16	16	16
57	1749	15	15	15
58	c.1750	16		
59	1750	20		
60	1751	16	16	16
61	1752	16	16	
62	1753	16½	16½	
63	1755	17	17	17
64	1756	17	17	17
65	1758	16	16	
66	1758	18	18	18
67	1760	17	17	17

68	1762	17	17	17
69	1764	17	17	17
70	1765	16	16	
71	1765	17	17	17
72	1765	18	18	18
74	1775	18	18	18
75	1775	18	18	
76	*c.*1775	18		
77	1776	17	17	17
78	1777	17	17	17
79	1777	18	18	18
80	1778	19	19	19
81	1779	19	19	19
82	1780	20	20	20
83	*c.*1780	19		
84	1783	18	18	
85	1787	16	16	
86	1788	18	18	18
87	1788	18	18	18
88	1789	17	17	
89	1790	18	18	
90	1790	17	17	
91	1791	17	17	17
92	1793	16	16	16
95	1805	14	14	

Widths given in some contemporary navigational manuals are listed below.

1/2 inch	Coignet 1580	12,7	mm
3/4 inch	Blundeville 1594	19	mm
1/2 inch	De Graaf 1658	12,7	mm
6-7 lignes	Bouguer 1698	13,5-15,7	mm
1/2-3/4 inch	De Vries 1702	12,7-19	mm

Appendix VII

Stamp patterns found on surviving Van Keulen staffs, on either side of their initials and the year of manufacture. An elevated digit indicates more surviving staffs for that particular year.

Pattern	Years
* * * *	1691, 1700, 1701, 1706
* * *	1702, 1706, 1733, 1745, 1749, 1756
*	1710, *c.* 1710, 1717, 1721, 1736
* *	1724
* * *	1740, 1749, 1751, 1755, 1758
* * *	1760, 1762, 1765^{2}
* * *	1765, 1775
* * *	1776, 1777^{2}
* * *	1778, 1779, 1780, 1788^{2}, 1791, 1793

Abbreviations

ARA	Algemeen Rijksarchief, The Hague
GAA	Gemeentearchief, Amsterdam
NA	Not available
NSA	Nederlands Scheepvaartmuseum, Amsterdam
UMU	Universiteitsmuseum, Utrecht
VOC	Verenigde Oostindische Compagnie

Bibliography

Manuscripts and non printed sources

Ms NSA 'Alphabetical list of Dutch Instrumentmakers compiled by W.F.J. Mörzer Bruyns 1970-1986 from several sources', Amsterdam, 1986.
Ms NSA 'The Blankman genealogy. The decendants in Holland and America of Pieter Blankman of North Holland *c.* 1680'. Compiled by Donald Warren Blankman, 1979. (with 1985 supplement).
Ms UMU 'Catalogus der Sterrekundige Werktuigen'. Compiled by R. van Rees, Utrecht, November 1839.
Ms UMU 'Catalogus der Sterrekundige Werktuigen'. Compiled by R. van Rees, Utrecht, July 1843.
Ms UMU 'Catalogus der sterrekundige werktuigen, gereedschappen, meubelement en boeken behoorende aan de Sterrewacht der Hoogeschool te Utrecht', *c.* 1884.
Ms UMU 'Catalogus der sterrekundige werktuigen, gereedschappen, meubelement en boeken behoorende aan de Sterrewacht der Hoogeschool te Utrecht', *c.* 1886.
Ms UMU 'Inventaris voor 1937 '[Inventory of Utrecht Observatory], Utrecht, *c.* 1937.
Ms NSA 'Kasboek betrekking hebbend op het zeevaartschooltje van Pieter Holm te Amsterdam, 1743-1774'.
Ms NSA 'Lijst van instrumenten in de verzameling van het Nederlandsch Historisch Scheepvaart Museum te Amsterdam'. Compiled by W.F.J. Mörzer Bruyns, 1971.

Works contemporary to the Cross-Staff

Cornelis Anthonisz. *Onderwijsinge vander zee / om Stuermanschap te leeren*. Amsterdam, 1558.
[Girolamo Alberti], *Introduzione all arte nautica: per uso de piloti, e capitani di nave, e per il migliore servizio de commandanti spora il mare*. Venice, 1715.
Petrus Apianus, *Cosmographie*. Antwerp, 1553.
James Atkinson, *Epitome Of The Art of Navigation Or A Short and Easy Way to become a Compleat Navigator*. London, 1718.

Nicolas Bion, *Traité de la construction et des principaux usages des instruments de mathematique*. La Haye, 1722.
Willem Jansz. [Blaeu], *The Light of Navigation*. Amsterdam, 1612.
Thomas Blundeville, *His Exercises, containing six Treatises*. London, 1594.
Jean Bouguer, *Traité Complet de la Navigation*. Paris, 1698.
William Bourne, *A Regiment for the Sea. ..., 1574*, in E.G.R. Taylor (ed.) *William Bourne, A Regiment for the Sea and of his writings on navigation*. Hakluyt Society second series no. CXXI, Cambridge, 1963.
Joost van Breen, *Stiermans Gemack, ofte een korte Beschryvinge vande Konst der Stierlieden*. The Hague, 1662.

Michiel Coignet, *Nieuwe Onderwijsinghe op de Principaelste Puncten der Zeevaert*. Antwerp, 1580.
Jacob Aertsz. Colom, *The Fierrie Sea-Columne*. Amsterdam, 1633.

Jacob Aertsz. Colom, *De Lichtende Colomne ofte Zee-Spiegel*. Amsterdam, 1652.
Nathaniel Colson, *The Mariners New Kalender. Containing.....the description and use of the ....forestaff..etc.* London, 1685.
Martin Cortes, *Breve Compendio de la Sphera y de la arte de Navegar*. Sevilla, 1551.

John Davis, *The Seamans Secrets*. London, 1607, in: Albert Hastings Markham (ed.) *The Voyages and Works of John Davis the Navigator*. Hakluyt Society, London, 1880.
Ezechiel de Decker, *Practyck vande Groote Zee-vaert*. Rotterdam, 1659.
Hendrik Doncker, *De Lichtende Columne ofte Zee-spiegel*. Amsterdam 1655.
Robert Dudley, *Arcano del mar; diviso in libri sei..*, 2nd ed. 2 vls, Florence, 1661.

Georges Fournier, *Hydrographie Contenant la Theorie et la Practique de Toutes les Parties de la Navigation*. Paris, 1643.
Gemma Frisius, *De Principiis Astronomiae & Cosmographiae*. Louvain, 1530.

Andrés Garcia de Cespedes, *Regimiento de Navegacion*. Madrid 1606.
Henry Gellibrand, *An Epitome of Navigation*. London, 1695.
Claes Hendricksz. Gietermaker, *'t Vergulde Licht Der Zee-vaert, Ofte Konst der Stuerlieden*. Amsterdam, 1660.
Claes Hendricksz. Gietermaker, *'t Vergulde Licht Der Zee-vaert, Ofte Konst der Stuerlieden*. Amsterdam, 1742.
Pieter Goos, *The Lighting Columne or Sea Mirrour*. Amsterdam, 1662.
Abraham de Graaf, *De Seven Boecken Van de Groote Zeevaert*. Amsterdam, 1658.
Edmund Gunter, *The Description And Use of The Sector, Crosse-staffe and other Instruments*. London, 1636.

Aelbert Haeyen, *Een corte onderrichtinge belanghende die kunst vander Zeevaert*. Amsterdam, 1600.
Joseph Harris, *A Treatise of Navigation*. London, 1730.

Thomas Hood, *The use of the two Mathematicall instruments, the crosse Staffe (differing from that in common use with mariners) And the Iacob's staffe*. London, 1590.

Dirk Kruik, *Gronden der Navigatie Ofte Stuurmans Konst*. Rotterdam, 1737.

Cornelis Jansz. Lastman, *De Schat-Kamer des grooten seevarts-kunst*. Amsterdam, 1653(?).
Cornelis Jansz. Lastman, *Kunst der Stuerluyden*. Amsterdam, 1657.

Dirck Makreel, *De Lichtende Leyd-starre der Groote Zee-vaert*. Amsterdam, 1671.
Pedro de Medina, *Arte de navegar*. Valladolid, 1545.
Pedro de Medina, *L'Art de Naviguer*. Lyon, 1554.
Pedro de Medina, *Regimento de Navegacion*. Sevilla, 1552.
Pedro de Medina, *De Zeevaert oft Conste van ter Zee te varen*. Antwerp, 1580.
Adriaan Metius, *Nieuwe Geographische Onderwysinge*. Amsterdam, 1621.
Claude-Francois Millet Dechales, *L'Art de Naviger*. Paris, 1677.
John Hamilton Moore, *The Practical Navigator and Seaman's New Daily Assistant*. London, 1772.
Jonas Moore, *A New Systeme of the Mathematicks*. London, 1681.

Pedro Nuñez, *De Arte Atque Ratione Navigandi*. Coimbra, 1573.

Simon Pietersz., *Stuermans Schoole*. Amsterdam, 1659.

John Robertson, *The Elements of Navigation; Containing the Theory and Practice*. London, 1754.
Werner von Rosenfeldt, *Navigation en Eller Styrmanskonsten*. Stockholm, 1693.

John Seller, *Practical Navigation*. London, 1689.
Antoni Struick, *Verhandeling over de Zeevaartkunde*. Rotterdam, 1768.

Verhandeling over de inrichting en het gebruik der octanten en sextanten van Hadley. Amsterdam, 1788.
Claas Jansz. Vooght, *De Zeemans Wegh-wyser.* Amsterdam, 1706.
Klaas de Vries, *Schat-kamer ofte Konst der Stier-lieden.* Amsterdam, 1702.

Lucas Jansz. Waghenaer, *Spieghel der Zeevaerdt.* Leiden, 1584.
Lucas Jansz. Waghenaer, *Thresoor der zeevaert.* Leiden, 1592.
Andrew Wakely, *The Mariner's Compass Rectified, with the Use of all Instruments of Navigation.* London, 1664.
Ioannis Verneri [Johann Werner], *Nova translatio primi libri geographiæ cl. Ptolæmæi paraphrasim prooemium.* Nurnberg, 1514.
Govaert Willemsen van Hollesloot, *Die caerte vande Oost ende West zee.* Harlingen, 1594.

Rodrigo Zamorano, *Cort Onderwijs vande Conste der Seevaert.* Amsterdam, 1598.

Secondary works

Baddeley, Jon *Nautical Antiques & Collectables.* London, 1993.
Bedini, Silvio A. *Thinkers and Tinkers. Early American Men of Science.* New York, 1975.
Bedini, Silvio A 'At the Sign of the Compass and Quadrant. The Life and Times of Anthony Lamb'. *Transactions of the American Philosophical Society*, 74, 1 (1984) pp.1-84.
Berghaus, A. 'Ein historischer Ruckblick'. *Central-Zeitung für Optik and Mechanik*, 14 (1893), pp.253-256 & 266-268.
Bos, J.M. 'Vragenrubriek'. *Antiek*, 4 (1970) p.434.
Breusing, Arthur 'Zür Geschichte der Geographie: 2 Regiomontanus, Martin Behaim under der Jakobsstab'. *Zeitschrift der Gesellschaft für Erdkunde*, Berlin, 4 (1869) pp.97-106.
Brewington, M.V. 'Notes on the cross-staff'. *The American Neptune*, XIV (1954) pp.187-191.
Brewington, M.V. 'Four American cross-staves'. *The American Neptune*, XIX (1959) pp.138-140.
Brewington, M.V. *The Peabody Museum collection of navigating instruments with notes on their makers.* Salem, 1963.
Brinckmann, J. *Museum für Kunst und Gewerbe im Hamburg. Bericht für das Jahr 1895 vom Director Professor Dr. Justus Brinckmann.* Hamburg, 1896.
Bryden, D.J. *Scottish Scientific Instrument-Makers.* Edinburgh, 1972.
Byzondere veiling betreffende scheepvaart en handel, bevattende o.a. het winkel- en magazijn meubilair van de oude firma Van Keulen ... in het verkooplokaal van H.G. Bom ... 7 October 1919. Amsterdam, 1919.

Catalogus van eene fraaije verzameling van Nederduitsche, Latynsche, Fransche, Engelsche en Hoogduitsche meest net geconditioneerde Boeken,...enz. Amsterdam, 1819.
Catalogus der voorwerpen betreffende scheepvaart en scheepsbouw, ingezonden op de Tentoonstelling der Kon. Ned. Zeil- en Roeivereeniging. Amsterdam, 1853. Verslag der Commissie ter beoordeling van de voorwerpen betreffende scheepvaart en scheepsbouw ... enz. Amsterdam, 1854.
Catalogus der belangrijke bibliotheken ... nagelaten door wijlen ... Jacob Swart ... alsmeede van het ouds vermaarde boeken-, kaarten- en instrumenten magazijn van de firma Wed. Gerard Hulst van Keulen te Amsterdam ... in het verkooplokaal van H.G. Bom. Amsterdam, 1885.
Catalogus van de geschiedkundige tentoonstelling van het Nederlandsche zeewezen. The Hague, 1900.
Catalogus der tentoonstelling van landkaarten, volkenkunde en scheepvaart, Antwerpen, Mei-Juli 1902, ingericht door het Kon. Aardrijkskundig Genootschap te Antwerpen. Antwerp, 1902.
Catalogue Exposition Societé Belge d'Astronomique, 1945.

Christian IV and Europe. The 19th Art Exhibition of the Council of Europe. Nationalmuseet. Copenhagen, 1988.
Clifton, Gloria C. *Directory of British Scientific Instrument Makers 1550-1851*. forthcoming, London, 1995.
Cotter, C.H. *A history of nautical astronomy*. London, 1968.
Cotter, C.H. *A history of the navigator's sextant*. Glasgow, 1983.
Cox, G.A. 'De jacobsstaf of de gemeyne graedboghe'. *Jaarverslag Vereeniging Nederlandsch Historisch Scheepvaart Museum 1952-53*, pp.35-48.
Crone, E. *Cornelis Douwes 1712-1773. Zijn leven en werk*. Haarlem, 1941.
Crone, E. 'De vondst op Nova Zembla. Een hernieuwd onderzoek der navigatie-instrumenten'. *Bulletin van het Rijksmuseum*, 14 (1966) pp.71-85.
Curtze, M. 'Die Abhandlung des Levi ben Gerson uber Trigonometrie und den Jacobsstab'. *Bibliotheca Mathematica*, 12 (1898) pp.97-112.

Daumas, Maurice *Scientific Instruments of the 17th & 18th Centuries and their Makers*. London, 1972.
Davids, C.A. *Zeewezen en wetenschap. De wetenschap en de ontwikkeling van de navigatietechniek in Nederland tussen 1595 en 1815*. Amsterdam/Dieren, 1986.
Davids, C.A. 'On the diffusion of nautical knowledge from the Netherlands to northeastern Europe, 1550-1850'. *From Dunkirk to Danzig. Essays in honour of prof. dr. J.A. Faber*. Amsterdamse Historische Reeks, Grote serie, V. Hilversum, 1988, pp. 217-236.
Davids, C.A., J.A. van der Veen & E.A. de Vries 'Van Lastman naar Gietermaker. Claes Hendricksz. Gietermaker (1621-1667) en zijn leerboeken voor de stuurmanskunst'. *Tijdschrift voor zeegeschiedenis*, 8 (1989) pp.149-177.
Deiman, J.C. 'Het instrumentarium van de Utrechtse sterrewacht'. *Het instrument in de wetenschap. Tijdschrift voor de geschiedenis der geneeskunde, natuurwetenschappen, wiskunde en techniek*, 10 (1987) pp.174-189.
Discovery 1778. Captain James Cook and the Peoples of the Pacific. Vancouver, 1978.
Donkersloot-de Vrij, Marijke *Drie generaties Blaeu. Amsterdamse cartografie en boekdrukkunst in de zeventiende eeuw*. Zutphen, 1992.
Dreier, Franz Adrian *Winkelmessinstrumente Vom 16. bis zum frühen 19. Jahrhundert*. Berlin, 1979.

Eckhardt, W. & K. *Wissenschaftliche Instrumente, bearbeitet von Wolfgang und Karin Eckhardt, Kataloge des Museums für Kunst und Gewerbe Hamburg*. forthcoming, Hamburg.
Ericsson, Christoffer H. 'Borstö'. *The International Journal of Nautical Archaeology*, 1 (1972) p.191.

Feyfer, H.J. de 'De graadstok of jacobsstaf'. *Jaarverslag 'Het Admiraliteitshuis' 1971*, pp.10-18.
Focus Behaim Globus, 2 vls., Germanisches Nationalmuseum, Nurnberg, 1992.
Forty, Gerald 'Sources of Latitude Error in English Sixteenth Century Navigation'. *The Journal of Navigation*, 36 (1983) pp.388-402.
Franco, Salvador Garcia *Instrumentos Nauticos en el Museo Naval*. Madrid, 1959.
Fries Scheepvaart Museum Gids. Sneek, s.a.

Gelcich, E. 'Beiträge zur Geschichte der mathematischen Geographie. a) Zur Geschichte des Jakobsstabes'. *Das Ausland*, 66 (1893) pp.337-339.
Gernez, D. 'Le libraire neerlandais Joannes Loots et sa maison d'editions maritimes'. *Mededelingen Academie van Marine van België*, 4, VIII (1954) pp.23-69.
Giordano, Raymond V. 'The Holyoke instruments at Harvard'. *Rittenhouse*, (8) 2, (1994) pp.33-40.
Günther, S. 'Die Erfindung des Baculus Geometricus'. *Bibliotheca Mathematica*, 1st Ser. (1885) pp.137-140.
Günther, S. 'Die erste Anwendung des Jakobsstabes zur geographischen Ortsbestimmung'. *Bibliotheca Mathematica*, 4 (1890) pp.73-80.

Günther, S. 'Uber den Jakobsstab als Hauptinstrument der geographischen Ortsbestimmung in fruherer Zeit'. *Verhandlung der Gesellschaft Deutscher Naturforscher und artze*, 67, Versammlung zu Lübeck. 16-20. September 1895, II Vol. 1 (Leipzig 1896) pp.142-143.
Günther, S. 'Der Jakobsstab als Hilfsmittel geographischer Ortsbestimmung'. *Geographische Zeitschrift*, 4 (1898) pp.157-167. Reprinted in Wolfgang Köberer (ed) *Das rechte Fundament der Seefahrt*. Hamburg, 1982.

Haan, H. de 'Graadstokken of jakobsstaven in Friese Musea'. *Fries Museumbulletin*, (December 1993), pp.8-11.
Harriet Wynter Ltd. Arts & Sciences June Catalogue 1983. London, 1983.
Harriet Wynter Ltd. Arts & Sciences Commemorative Catalogue 1986. London, 1986.
Heywood, H. 'A Brief Description of the 18th Century Scientific Instruments recently discovered at Longleat, the home of the Marquess of Bath'. *Bulletin of the British Society for the History of Science*, 1, no.10 (May 1954) p.246.
Holbrook, Mary *Science Preserved. A directory of scientific instruments in collections in the United Kingdom and Eire*. London, 1992.
Hollandia Compendium. A contribution to the history, archaeology, classification and lexicography of a 150 ft Dutch East Indiaman (1740-1750). Jerzy Gawronski (ed.) Amsterdam, 1992.
Horsburgh, Ellice M. 'The Cross-Staff and its use in Navigation'. *Scottish Geographical Magazine*, 46 (1930) pp.92-100.
Horsky, Zdenek & Otilie Skopová *Astronomy Gnomonics. A catalogue of the 15th to the 19th centuries in the collection of the National Museum Prague*. Prague, 1968.
Howse, Derek *Greenwich Time and the Discovery of Longitude*. Oxford, 1980.

Inventaire des instruments scientifiques historiques conserves en Belgique. 2 vls. Centre National d'Historique des Sciences. Brussels, 1959 & 1960.
An Inventory of the Navigation and Astronomy Collections in the National Maritime Museum Greenwich. 3 vls. London, 1970-1982.

Jaspers, H. 'Over de constructie en het gebruik van de jacobsstaf'. *Marineblad*, (1968) pp.491-500.
Jonge, J.K.J. de *De opkomst van het Nederlandsch gezag in Oost-Indië; verzameling van onuitgegeven stukken uit het oud-koloniaal archief*. dl. 1, (1595-1610). The Hague, 1862.
Jonge, J.K.J. de *Nova Zembla; verslag over de voorwerpen door Nederlandsche zeevaarders op Nowaja Semlja bij hun vertrek in 1597 achtergelaten en in 1876 door Ch. Gardiner Esq. aldaar teruggevonden*. The Hague, 1877.

Kaijzer, I. *Vraked vid Alvsnabben. Dokumentation, last och utrusting. Rapport 13 Statens Sjöhistoriska Museum*. Stockholm, s.a. [1982].
Keulen, E.O. van, W.F.J. Mörzer Bruyns & E.K. Spits *In de Gekroonde Lootsman. Het kaarten-, boekuitgevers en instrumentmakershuis Van Keulen te Amsterdam 1680-1885*. Utrecht, 1989.
Knöblich, Eberhard 'Jacobsstab'. *Mass, Zahl und Gewicht*. Wolfenbüttel, 1989.
Köberer (ed.), Wolfgang *Das rechte Fundament der Seefahrt*. Hamburg, 1982.
Koeman, C. *Atlantes Neerlandici*. 5 vls, Amsterdam, 1967-1971.
Kok, M. 'Cartografie van de firma Van Keulen 1674-1880'. *'In de Gekroonde Lootsman'. Het kaarten-, boekuitgevers en instrumentenmakershuis Van Keulen te Amsterdam 1680-1885*. Utrecht, 1989, pp.15-43.
Krogt, P.C.J. van der *Advertenties voor kaarten, atlassen, globes e.d. in Amsterdamse kranten 1621-1811*. Utrecht, 1985.

Langenbach, M. *Onbekend talent. Leerlingen van de Utrechtse Fundatie van Renswoude*. Zutphen, 1991.
Ledeboer, A.M. *Alphabetische Lijst der Boekdrukkers, Boekverkoopers en Uitgevers in Noord-Nederland*. Utrecht, 1876.

Losman, Arne & Irene Sigurdsson 'Äldre vetenskapliga instrument på Skokloster. Scientific Instruments of the 16th-18th Centuries at Skokloster Castle'. *Skokloster Studies*, 10, (1975) pp.86-103.

Maddison, Francis *Medieval Scientific Instruments and the Development of Navigational Instruments in the XVth and XVIth Centuries*. Agrupamento de Estudos de Cartografia Antiga, Coimbra 1969.
Mailly, N.E. *Catalogue du Musée de l'Industrie*. Brussels, 1846.
Meskens, A. 'Michiel Coignet's Nautical Instruction'. *The Mariner's Mirror*, 78, 3 (1992) pp.257-276.
Minow, Helmut *Historische Vermessungsinstrumente. Ein Verzeichnis der Sammlungen in Europa*. Wiesbaden, 1990.
Molaug, Svein & Rolf Scheen *Fregatten 'Lossen'. Et kulturhistorisk skattkammer*. Norsk Sjøfartsmuseum, Skrift no.43, Oslo, 1983.
Mörzer Bruyns, W.F.J. 'De navigatie-instrumenten geborgen uit de Hollandia (vergaan in 1743)'. *Jaarverslag Vereeniging Nederlandsch Historisch Scheepvaart Museum 1974*, pp.44-48.
Mörzer Bruyns, W.F.J. 'Navigatie-instrumenten van de zeebodem 16e tot 19e eeuw'. *Het instrument in de wetenschap. Tijdschrift voor de geschiedenis der geneeskunde, natuurwetenschappen, wiskunde en techniek*, 10 (1987) pp.263-282.
Mörzer Bruyns, W.F.J. 'De instrumenten van de firma Van Keulen'. *'In de Gekroonde Lootsman'. Het kaarten-, boekuitgevers en instrumentenmakershuis Van Keulen te Amsterdam 1680-1885*, Utrecht, 1989, pp.61-71.
Mörzer Bruyns, W.F.J. 'More Information on the Cross Staff'. *Bulletin of the Scientific Instrument Society*, 25 (1990) p.15.
Mörzer Bruyns, W.F.J. 'Navigation on Dutch East India Company Ships around the 1740s'. *The Mariner's Mirror*, 78, 2 (1992) pp.143-154.
Navigationsskolorna 1841-1941. Utställning med anledning av 100-årsjubileet anordnad av Sjöfartsmuseet Göteborg. Gothenburg, 1941.
Nierop, L. van 'Naamregister van alle voornaamste winkeliers der Stad Amstelredam (1767)'. *Jaarboek van het Genootschap Amstelodamum*, 29 (1932) pp.159-216.

Oldewelt, W.F.H. *Kohier van de personele quotisatie te Amsterdam over het jaar 1742*. 2 vls, Amsterdam, 1945.

Peters, Hubert J.M.W. *The Crone Library. Books on the Art of Navigation left by Dr. Ernst Crone to the Scheepvaart Museum in 1975 and books on the same subject acquired by the Museum previously*. Amsterdam, 1989.

Quetelet, A. *Annales de l'Observatoire Royal de Bruxelles*. Brussels, 1857.

Roche, John J. 'The radius astronomicus in England'. *Annals of Science*, 38 (1981) pp.1-32.
Rooseboom, Maria *Bijdrage tot de geschiedenis van de instrumentmakerskunst in de Noordelijke Nederlanden tot omstreeks 1840*. Leiden, 1950.

Samso, Julio *Instrumentos Astronomicos en la Espana Medieval. Su influencia en Europa*. Barcelona, s.a.
Schilder, Günter 'Organization and Evolution of the Dutch East India Company's Hydrographic Office in the Seventeenth Century'. *Imago Mundi. The Journal of the International Society for the History of Cartography*, 28 (1976) pp.61-78.
Schmidt, Friedrich 'Urformen eines alten nautischen Instrumentes (des Jakobsstabes)'. *Pfalzische Museum-Pfalzische Heimatkunde*, 43 (1926) pp.123-126.
Schück, A. 'Der Jakobsstab bei den Arabern'. *Die Natur*, 40 (1891) pp.352-355.
Schück, A. 'Der Jakobsstab'. *Jahresbericht der Geographischen Gesellschaft in München*, 16 (1894/95) pp.93-174.

Schück, A. 'Zur Entwicklung des Jakobsstabes'. *Mittheilungen der Geographischen Gesellschaft in München*, 8 (1913) pp.1-7.
Schwindrazheim, Hildemarie 'Der Jakobsstab. Ein bericht aus der Arbeit des Altonaer Museums'. *Altonaische Zeitschrift für Geschichte und Heimatkunde*, 4 (1935) pp.107-111.
Seefahrt einst und heute. Ausstellung in Stadtischen Museum Flensburg vom 11.6 bis 15.8 1976 in Verbindung mit der Industrie und Handelskammer zu Flensburg. Flensburg, 1971.
Simpson, Allen 'A Seventeenth Century Cross Staff'. *Bulletin of the Scientific Instrument Society*, 27 (1990) pp.8-9.
Steinschneider, Moritz 'Levi ben Gerson und der Baculus Jacobi'. *Bibliotheca Mathematica*, 2nd serie, 4 (1890) p.107.
Stimson, Alan Neal & Christopher St John Hume Daniel *The Cross Staff. Historic Development and Modern Use*. London, 1977.
Stimson, Alan *The Mariner's Astrolabe. A survey of known, surviving sea astrolabes*. Utrecht, 1988.
Suter, H. 'Zur Geschichte des Jakobsstabes'. *Bibliotheca Mathematica*, 2nd series, 9 (1895) pp.13-18.
Suter, H. 'Nochmals der Jakobsstab'. *Bibliotheca Mathematica*, 2nd series, 10 (1896) pp.13-15.

Taylor, E.G.R. *The Mathematical Practitioners of Hanoverian England 1714-1840*. Cambridge, 1966.
Taylor, E.G.R. *The Mathematical Practitioners of Tudor & Stuart England*. Cambridge, 1970.
Thirslund, Søren 'Samlingen af nautiske instrumenter på Handels-og Søfartsmuseet'. *Årbog 1984, Handels- og Søfartsmuseet på Kronborg*, Helsingør, 1984.

Warner, Deborah Jean 'Davis' Quadrants in America'. *Rittenhouse*, (3) 1, (1988) pp.23-40.
Waters, David W. *The Art of Navigation in England in Elisabethan and Early Stuart Times*. London, 1958.
Waters, D.W. *Science and the Techniques of Navigation in the Renaissance*. London, 1976.
Wolf, R. 'No 185'. *Astronomische Mittheilungen*, XXXXI, Zürich (1876) pp.23-24.
Wijk, W.E. van 'De Jacobsstaf', *Natuur en Mens*, 57 (1937) pp.40-42.
W[ijk], W.E. van 'Nog eens de Jacobsstaf', *Natuur en Mens*, 57 (1937) p.54.
Wynter, Harriet 'Another 17th Century Cross-staff'. *Bulletin Scientific Instrument Society*, 28 (1991) p.24.

Zijlstra, Jan 'Jakobsstaven in het Fries Museum'. *Fries Museumbulletin*, (June 1993) pp.3-4.
Zinner, Ernst *Deutsche und Niederländische Astronomische Instrumente des 11.-18. Jahrhunderts*. Munich, 1956.

References

An Introduction to the History of Astronomical Navigation and Navigational Instruments

1. Before that time, seafarers from the Middle East and from Scandinavian countries had developed navigational skills which enabled them to sail long distances from their home ports, out of sight of land. There is, however, no evidence that they made use of methods of astronomical navigation. Waters, *Science and the Techniques*, pp.1-2.
2. Longitude is the numerical value of the arc of the equator, between a standard meridian and the meridian of the observer. Most countries had their own standard meridian (Paris, London, Amsterdam, Teneriffe). Since 1884 the meridian over Greenwich is internationally accepted as the standard meridian.
3. Verneri *Nova translatio*. See Howse, *Greenwich Time*, pp.6-9.
4. Waters, *Science and the Techniques*, p.4.
5. Howse, *Greenwich Time*, pp.10-13.
6. The log, for measuring the speed through water, was generally introduced at sea in the 1590s. Waters, *The Art of Navigation*, p.122.
7. Waters, *Science and the Techniques*, p.29.
8. Waters, *The Art of Navigation*, p.46n.
9. Stimson, *The Mariner's Astrolabe*, p.16.
10. The declination of a celestial object is the numerical value of the arc of a celestial meridian intercepted between the object and the equinoctial. The maximum declination of the sun is $23^1/_2°$ North or South.
11. Waters, *The Art of Navigation*, p.52. Declination tables of the sun were available in manuscript as early as 1485. See Forty, 'Sources of Latitude Error', p.400.
12. Zenith distance is the complement of the altitude. In the formula, North is + and South is -. Thus, Northern latitude, Northern declination and the celestial body observed in the North are +, etc.
13. Waters, *The Art of Navigation*, pp.43-47. The exact distance was a matter of controversy in the sixteenth century. Using the wrong figure could cause an error in the calculation, see Forty, 'Sources of Latitude Error', p.400.
14. Waters, *Science and the Techniques*, p.14n.
15. Ibid., p.14.
16. In Amsterdam, Zeeland, Rotterdam, Delft, Enkhuizen and Hoorn.
17. Schilder, 'Organization and Evolution', p.61.
18. Ibid., pp.63-65 and Mörzer Bruyns, 'Navigation', pp.143-144.
19. Davids, *Zeewezen en wetenschap*, p.398.
20. Schilder, 'Organization and Evolution', pp.63-65 and Mörzer Bruyns, 'Navigation', p.144.
21. Davids, *Zeewezen en wetenschap*, pp.172-173. An exception were ships of the Zeeland Chamber of the VOC on board which the *spieghelboog* was used. This will be dealt with in a later chapter.
22. Ibid., pp.226-231.
23. Ibid., pp.229-230. After 1762 the chief mate also received an octant.
24. Crone, *Cornelis Douwes* and Cotter, *A history of nautical astronomy*, pp.148-149.
25. Crone, *Cornelis Douwes*, pp.263-290.
26. Cotter, *A history of the navigator's sextant*, pp.137-138.
27. Ibid., pp.145-147.

Historiography and Sources on the Cross-Staff

1. Köberer, *Das rechte Fundament*, pp.11-12.
2. Ibid., p.296.
3. Breusing, 'Zur Geschichte der Geographie'. Reprinted in Köberer, *Das rechte Fundament*, pp.186-193.
4. Köberer, *Das rechte Fundament*, p.298.
5. Schück, 'Der Jakobsstab'.
6. Schwindrazheim, 'Der Jakobsstab' and Brewington, 'Four American cross-staves'.
7. Cox, 'De jacobsstaf'.
8. Cotter, *A history of the navigator's sextant.*
9. Stimson & Daniel, *The Cross-Staff.*
10. Forty, 'Sources of Lattitude Error'.
11. Roche, 'The radius astronomicus'.
12. Koeman, *Atlantes Neerlandici.*
13. Peters, *The Crone Library.*
14. At the Project for Historical Bibliography in Newcastle-upon-Tyne.
15. Verneri, *Nova translatio*, Apianus, *Cosmographie* and Frisius, *De Principiis.*

16. Cortes, *Breve Compendio.*
17. De Medina, *Arte de Navegar* and Nunes, *De Arte.*
18. Coignet, *Nieuwe Onderwijsinghe*, p.14, Fournier, *Hydrographie*, p.495 and De Graaf, *De Seven Boecken*, p.4.
19. Davids, *Zeewezen en wetenschap*, p.57.
20. Coignet, *Nieuwe Onderwijsinghe*, Waghenaer, *Spieghel* and Davis, *The Seamans Secrets.*
21. Pietersz., *Stuermans Schoole*, p.135.
22. Schilder, 'Organization and Evolution', also Mörzer Bruyns, 'Navigation'.
23. Zamorano, *Cort Onderwijs.*
24. Mörzer Bruyns, 'Navigatie-instrumenten', pp. 144-146.

Introducing the Cross-Staff

1. Gellibrand, *An Epitome of Navigation.* The first edition, of which no copy is known to survive, probably dates from *c.* 1634, see Taylor, *The Mathematical Practitioners*, p.349.
2. After the Latin *arcuballista.*
3. Cox, 'De jacobsstaf', pp.38-40.
4. The Dutch, German, Swedish and Russian name translated into English means 'degree-stick' or 'degree-bow'.
5. Hood, *The use of the two Mathematicall instruments*, p. cii verso.
6. Cox, 'De jacobsstaf', p.38, refers to Genesis 28:12 (King James Bible) where it says 'And he [Jacob] dreamed, and behold a ladder set up on the earth, and the top of it reached to heaven: and behold the angels of God ascending and descending on it'.
7. Van Wijk, 'De Jacobsstaf', p.40, refers to Genesis 30:37 (King James Bible) where it says 'And Jacob took him rods of green poplar, and of the hazel and chestnut tree; and pilled white strakes in them, and made the white appear which *was* in the rods'.
8. Günther, 'Die erste Anwendung des Jakobsstabes', pp.73-74 and Schück, 'Der Jakobsstab'. About the possible Jewish-Arabian origin of the baculus Jacob see Steinschneider, 'Levi ben Gerson', Schück, 'Der Jackobsstab bei den Arabern', Maddison, *Medieval Scientific Instruments*, p.47 and Daumas, *Scientific Instruments*, p.10.
9. Günther, 'Die erste Anwendung des Jakobsstabes'. See also Breusing, 'Zür Geschichte der Geographie' and Schück, 'Der Jakobsstab'.
10. Cox, 'De jacobsstaf', p.39.
11. Schück, 'Der Jakobsstab', pp.131-133; Schmidt, 'Urformen'; Waters, *The Art of Navigation*, p.53; Maddison, *Medieval Scientific Instruments*, p.50; Waters, *Science and the Techniques*, p.13 and Cotter, *A history of the navigator's sextant*, p.65.
12. Maddison, *Medieval Scientific Instruments*, p.47 and Cox, 'De jacobsstaf', p.39.
13. Roche, 'The radius astronomicus', pp.1-2 and Cotter, *A history of the navigator's sextant*, pp.18-19. See also Maddison, *Medieval Scientific Instruments*, pp.46-51.
14. In the British Museum, London and the Museo Nacional de Ciencia y Technologia (Instituto de San Isidro), Madrid, see Maddison, *Medieval Scientific Instruments*, p.49 and Holbrook, *Science Preserved*, p.153.
15. In the Istituto e Museo di Storia della Scienza, Florence. They are of wood and gilded brass, not signed, but attributed to the German maker Christóph Schissler (*c.* 1531-1609). They were bought for the Medici collection in the early seventeenth century.
16. Cotter, *A history of the navigator's sextant*, p.91.
17. Ibid., p.98.
18. Warner, 'Davis' Quadrants'.
19. Davids, *Zeewezen en wetenschap*, pp.173-175.
20. Ibid., p.124.
21. Waters, *The Art of Navigation*, p.xxvii.
22. Dudley, *Arcano*, libro quinto, pp.15-17.
23. Millet Deschales, *L'Art de Naviger*, p.64.
24. Harris, *A Treatise*, pp.184-185.
25. Forty, 'Sources of Latitude Error', p.395.
26. Cotter, *A history of the navigator's sextant*, p.81.
27. Stimson & Daniel, *The Cross-Staff*, pp.17-18.
28. Ibid., p.19.
29. Moore, *A New Systeme*, p.241. Warner, 'Davis' Quadrants', p.23.
30. Blaeu, *The Light of Navigation*, chapter xiii and Colom, *De Lichtende Colomne.*
31. Pietersz., *Stuermans Schoole*, p.34, Moore, *A New Systeme*, p.239 and Colson, *The Mariners New Kalender*, p.81.
32. Gietermaker, *'t Vergulde Licht* 1660, p.54.
33. Forty, 'Sources of Latitude Error', pp.391-392.
34. Cotter, *A history of the navigator's sextant*, p.71.
35. Ibid., p.71. It was also condemned by Waghenaer, *Thresoor*, p.xviii and by Metius, *Nieuwe Geographische Onderwijsinge*, p.25.
36. Fournier, *Hydrographie*, pp.495-508.
37. Davis, *The Seamans Secrets*, p.263.
38. Forty, 'Sources of Latitude Error', p.392.
39. Harris, *A Treatise*, p.183.
40. Forty, 'Sources of Latitude Error', pp.393 & 395.
41. Waters, *The Art of Navigation*, p.585.
42. Forty, 'Sources of Latitude Error', p.400. When using quadrant and astrolabe, dip is avoided as the horizon is not used.

43. Coignet, *Nieuwe Onderwijsinghe*, pp.14-15verso. On Coignet's contribution to navigation see also Meskens, 'Michiel Coignet'.
44. Waters, *The Art of Navigation*, pp.204-205.
45. Ibid., p.171.
46. Colom, *The Fierrie Sea-Columne*, p.19.
47. De Graaf, *De Seven Boecken*, p.5.
48. Cox, 'De jacobsstaf', p.38.
49. Moore, *A New Systeme*, pp.242-243.
50. Davids, *Zeewezen en wetenschap*, p.403.
51. Makreel, *De Lichtende Leyd-starre*, pp.140-144.
52. Forty, 'Sources of Latitude Error', p.390.
53. Blaeu, *The Light of Navigation*, chapters xiii & xiiii and Metius, *Nieuwe Geographische Onderwijsinge*, pp.23-26.
54. Makreel, *De Lichtende Leyd-starre*, p.141.
55. Vooght, *De Zeemans Wegh-wyser*, pp.326-332.
56. Goos, *The Lighting Columne.*
57. Cortes, *Breve Compendio*, p.lxxxi and Coignet, *Nieuwe Onderwijsinghe*, pp.14-15verso.
58. Moore, *A New Systeme*, pp.241-242.
59. Goos mentioned 11,000 parts, Moore preferred 100,000, but, the vane being too short for such a division, both wrote that 1,000 parts, although less accurate, would also suffice.

Production and Diffusion of the Cross-Staff

1. Bourne, *A Regiment for the Sea*, p.207.
2. Hood, *The use of two Mathematical instruments.* See Appendix 1.
3. Haeyen, *Een corte onderrichtinge*, p.43.
4. Garcia de Gespedes, *Regimiento*, p.146.
5. Davids, *Zeewezen en wetenschap*, p.121.
6. Fournier, *Hydrographie*, p.504.
7. Colom, *De Lichtende Colom* & *The Fierre Sea-Column.* See also Schück, 'Der Jakobsstab', pp.156-158.
8. See Appendix I.
9. Von Rosenfeldt, *Navigation*, p.33.
10. Seller, *Practical Navigation*, p.171.
11. See Appendix III. Of many of the Dutch maker's names no cross-staff is known to exist. Considering the substantial production by the Van Keulens, it may be that some of these makers were working for the firm of Van Keulen, under the firm's name.
12. Mörzer Bruyns, 'De instrumenten', pp.61 & 64.
13. Waters, *The Art of Navigation*, p.79.
14. Ibid., p.502.
15. Davids, 'On the diffusion'.
16. Blundeville, *His Exercisis*, pp.313-314verso.
17. Gunter, *The Description And Use*, p.1.
18. Wakely, *The Mariner's Compass Rectified*, p.195.
19. Von Rosenfeldt, *Navigation*, p.33.
20. Appendix I and Bedini, *Thinkers and Tinkers*, pp.198 & 208.
21. Davids, *Zeewezen en wetenschap*, pp.121-122.
22. De Jonge, *De opkomst*, pp.182-183.
23. Davids, *Zeewezen en wetenschap*, p.121 and Appendix II.
24. Davids, *Zeewezen en wetenschap*, pp.171-172.
25. It had been used on some VOC ships as early as 1738, Ibid., pp.229-230.
26. Ibid., pp.231 & 352.
27. Ibid., pp.299-300.
28. Cotter, *A history of nautical astronomy*, p.72 and Daumas, *Scientific Instruments*, p.12. See also Davids, *Zeewezen en wetenschap*, p.174, who disaproved of Cotter's statement.
29. Robertson, *The Elements of Navigation*, pp.517 & 519.
30. Struick, *Verhandeling*, p.169.
31. Moore, *The Practical Navigator*, pp.40, 132 & 172.
32. *Verhandeling*, p.3.
33. Davids, *Zeewezen en wetenschap*, pp.231-232.

A Comparison Between Surviving Examples of the Cross-Staff and Written Sources

1. Cortes, *Breve Compendio*, p.lxxx, Millet Deschales, *L'Art de Naviger*, p.63, Bouguer, *Traité*, p.93, Bion, *Traité*, p.275 and Alberti, *Introduzione all arte nautica*, p.99.
2. Apianus, *Cosmographias Liber*, p.xvverso.
3. Maddison, *Medieval Scientific Instruments*, p.49.
4. Coignet, *Nieuwe Onderwijsinghe*, p.14, Zamorano, *Cort Onderwijs*, p.26 and Blundeville, *His Exercisis*, p.313.
5. Fournier, *Hydrographie*, p.498, advised 'well dried ebony'.
6. Ibid., p.498.
7. Davids, *Zeewezen en wetenschap*, p.172.
8. Seller, *Practical Navigation*, p.167.
9. Cortes, *Breve Compendio*, p.lxxx, De Graaf, *De Seven Boecken*, p.5, Seller, *Practical Navigation*, p.167 and Von Rosenfeldt, *Navigation*, p.33.
10. Bouguer, *Traité*, planche 4 and De Vries *Schatkamer*, pp.280-281.
11. Apianus, *Cosmographiais Liber*, p. xvverso, Cortes, *Breve Compendio*, p.lxxx and Zamorano, *Cort Onderwijs*, p.26 state the staff was a 'finger' thick.
12. As Danish and Schleswig-Holstein feet, Schück, 'Der Jakobsstab', p.97.
13. Cortes, *Breve Compendio*, p.lxxx and Zamorano, *Cort Onderwijs*, p.26.
14. Coignet, *Nieuwe Onderwijsinghe*, p.14, Fournier, *Hydrographie*, p.496, Bouguer, *Traité*, p.93 and Bion, *Traité*, p.275.
15. Blundeville, *His Exercises*, p.313, Moore, *A New Systeme*, p.237 and Robertson, *The Elements of Navigation*, p.519.

16. Alberti, *Introduzione all arte nautica*, p.99.
17. Von Rosenfeldt, *Navigation*, p.33.
18. De Graaf, *De Seven Boecken*, p.5.
19. Seller, *Practical Navigation*, p.167.
20. De Vries, *Schat-kamer*, p.62, soon followed by Bion, *Traité*, p.275.
21. In an appendix in Klaas de Vries, *Schat-kamer ofte Kunst der Stierlieden*, Amsterdam, 1777, in the library of the NSA.
22. De Vries, *Schat-kamer*, p.272.
23. Now in The Admiralty Board Room, Whitehall, London, and seen by the author in December 1993. Although a carving, the impression is that an original cross-staff was used as example. The end of the staff, with a width of 12 mm, is pyramid-shaped. The graduation is divided to 10' and the staff has three vanes.
24. When Harriet Wynter had her first replica cross-staffs made, the vanes were modelled after English examples, and made out of two pieces. When clamping the vane to the staff by turning the screw the two glued parts were forced apart, rendering the vane useless. This problem was solved as soon as she had vanes made out of one piece of wood. Personal communication to the author.
25. Coignet, *Nieuwe Onderwijsinghe*, p.14 and Blundeville, *His Exercises*, p.313.
26. As supplied to VOC ships in that year, see Appendix II.
27. Millet Deschales, *L'Art de Naviger*, p.63.
28. See Appendix II.
29. Gietermaker, *'t Vergulde Licht*, 1742, p.54.
30. See Appendix II.
31. Pietersz., *Stuermans Schoole*, p.34.
32. Seller, *Practical Navigation*, p.171. Brewington, *The Peabody Museum*, p.6, wrongly claimed the whole horizon vane was named Dutch shoe.
33. Seller, *Practical Navigation*, p.171.
34. Bourne, *A Regiment for the Sea*, p.209.
35. Colom, *De Lichtende Colomne* and Seller, *Practical Navigation*, p.171.
36. At five cents a *stuiver*. Advertised in an appendix in Klaas de Vries, *Schat-kamer ofte Kunst der Stierlieden*, Amsterdam, 1777, in the library of the NSA.
37. Crone, *Cornelis Douwes*, pp.46-47.
38. Moore, *A New Systeme*, p.237 and Colson, *The Mariner's New Kalender*, p.80.
39. Moore, *A New Systeme*, p.237 and Seller, *Practical Navigation*, p.167.
40. Colson, *The Mariners New Kalender*, p.81.
41. Coignet, *Nieuwe Onderwijsinghe*, p.16verso, Blundeville, *His Exercises*, p.314verso, Fournier, *Hydrographie*, p.503 and De Graaf, *De Seven Boecken*, p.5.
42. Moore, *A New Systeme*, p.237 and Colson, *The Mariners New Kalender*, p.80.
43. Coignet, *Nieuwe Onderwijsinghe*, p.14.
44. Moore, *A New Systeme*, p.237, Colson, *The Mariners New Kalender*, p.80 and Seller, *Practical Navigation*, p.167. The Italian Alberti, *Introduzione all arte nautica*, p.99, also gave the English sequence.
45. Zamorano, *Cort Onderwijs*, p.26.
46. Haeyen, *Een corte onderrichtinge*, pp.42-43.
47. De Vries, *Schat-kamer*, p.62.
48. Seller, *Practical Navigation*, p.167.
49. Harris, *A Treatise*, p.184.
50. Bourne, *A Regiment for the Sea*, p.207, Blundeville, *His Exercises*, p.313 and Seller, *Practical Navigation*, p.167.
51. Garcia de Cespedes, *Regimiento*, p.147. Although perhaps visible, because of the refraction it is indeed too low to give a reliable result.
52. Fournier, *Hydrographie*. De Graaf, *De Seven Boecken*, p.5 gives 2°. Moore, *A New Systeme*, p.237 and Seller, *Practical Navigation*, p.130, both give 3°.

Conclusion

1. See Stimson, *The Mariner's Astrolabe*.

The Checklist

1. Inv.2.271. Horsky & Skopová, *Astronomy*, pp.28 & 161. Letter from Dr Jan Mráz, director of the National Technical Museum in Prague dated 12/02/1988 to Mrs J. Schuller tot Peursum-Meijer, then at the University Museum in Groningen. According to Horsky & Skopová, this instrument was purchased in 1911 from Frederik Muller in Amsterdam. On 29 November 1911 a nineteenth-century copy of a cross-staff was auctioned as lot no.391 in the auction of the instruments of the Strozzi family. It was sold for *f* 75.
2. Zinner, *Astronomische Instrumente*, p.210. A staff of 1000 mm with graduation on all four sides and three vanes, 87, 295 and 515 mm long. Letter from Mrs Gudrun Heidle of the Grosse Kreisstadt, Lindau (Bodensee) 13 April 1990 to the author.
3. *Catalogus der voorwerpen*, nos.279 & 472-473.
4. When the effects of the firm of Van Keulen were auctioned there was no cross-staff among the numerous instruments, see *Catalogus der belangrijke bibliotheken* & *Byzondere veiling*.
5. A fragment of a ebony staff, 275 mm long and 6 x 6 mm wide. Three sides are in good condition, the fourth has carved marks at intervals of *c.* 2.4 cm. The maximun accuracy of the graduation is 30'. Judging by width and accuracy it may have been part of a mid-sixteenth-century cross-staff.

6. Handels- og Søfartsmuseet på Kronborg Helsingør, inv.K.3575 & K.3576. Sjöhistoriska Museet Stockholm, inv.S.666, S.1220 & S.1221. Norsk Sjøfartsmuseum Oslo, inv.2095, 2559, 8441, 8442, 8448 & 8455.
7. *Arbøg 1986. Handels- og Søfartsmuseet på Kronborg*, pp.24-25. Jens Kusk Jensen (1866-1939) was a Danish sea-captain and historian of navigation. The *Arbøg* also records an example by Jensen in the Science Museum, London.
8. Including a replica of the Tuttell staff, inv.S.160. Another copy of the same was used on board the replica of the *Golden Hinde* in 1975, inv.S.291 and there is a reconstruction commissioned by Harriet Wynter Ltd, inv.S.299. *An Inventory*, 1, section 12.
9. By Harold Ericson of Noank, inv.46.74.
10. Holbrook, *Science Preserved*, p.194 states that the Museum of the History of Science in Oxford owns cross-staffs, Professor G.L'E. Turner informed me that these are two replicas. Minow, *Historische Vermessungsinstrumente*, p.36 records a cross-staff in the Museum of Science and History in Birmingham. It is, in fact, a surveyor's cross. Minow, *Ibid.*, p.81 states that the Centralne Muzeum Morskie in Gdansk owns two cross-staffs; Dr Jerzy Litwin, the museum's deputy director, informed me that both are replicas.
11. Minow, *Historische Vermessungsinstrumente*, p.179.
12. Ibid., p.104.

Appendix I

1. Wolf, *Astronomische Mittheilungen*, pp.23-24.
2. Crone, 'De vondst op Nova Zembla', pp.75-76.
3. Ms GAA Poortersboek 17/149, 21 February 1730, son of Jan Janse Apegendam, distiller.
4. Schück, 'Der Jakobsstab', p.153 and Donkersloot-de Vrij, *Drie generaties Blaeu*, pp.22 & 25.
5. Ledeboer, *Alphabetische Lijst*, p.18. Davids, Van der Veen & De Vries, 'Van Lastman naar Gietermaker', pp.166-167.
6. Van Keulen, *In de Gekroonde Lootsman*, p.72.
7. Ms GAA Ondertrouwregister 478/230, 2 March 1658.
8. Clifton, *Directory*. Harris, *A Treatise*.
9. Ibid., and Taylor, *Mathematical Practitioners 1714-1840*, p.172.
10. Hood, *The use of the two Mathematical instruments*. Taylor, *Mathematical Practitioners*, p.189.
11. Ms GAA Ondertrouwregister 499/271, 23 September 1673.
12. Clifton, *Directory*.
13. Ms GAA Poortersboek 31/125, 24 January 1806.
14. Ms GAA Ondertrouwregister 744/123, 2 April 1768.
15. Ms GAA Poortersboek 25/90, 27 January 1769.
16. Ms GAA Ondertrouwregister 507/255, 18 May 1680.
17. Ms GAA Poortersboek 21/383, 24 January 1753, his father was seaman.
18. Ms GAA Ondertrouwregister 16 October 1648.
19. On the titlepage of Donker, *De Lichtende Columne*, the only known copy of which is in the British Museum, London. Koeman, *Atlantes Neerlandici*, 4, p.174.
20. Koeman, *Atlantes Neerlandici*, 4, pp.153-154. Ms NSA 'Alphabetical list', 1986.
21. Van der Krogt, *Advertenties*, nos 445 & 447.
22. Ms GAA Ondertrouwregister 473/427, 18 September 1654.
23. Van der Krogt, *Advertenties*, no.A88.
24. Ms GAA Ondertrouwregister 504/352, 3 December 1677.
25. Ms GAA Ondertrouwregister 542/209, 27 January 1708.
26. Schilder, 'Organization and Evolution', p.72.
27. Ms ARA VOC 7150. See Appendix II and III.
28. Ms GAA Ondertrouwregister 696/449, 11 March 1690.
29. Ms GAA Ondertrouwregister 691/119, 29 April 1678.
30. Clifton, *Directory*. Harris, *A Treatise*. Taylor, *Mathematical Practitioners 1714-1840*, p.161.
31. Van Nierop, 'Naamregister' and Rooseboom, *Bijdrage*, p.69. Ms NSA 'Alphabetical list', 1986.
32. Ms GAA Poortersboek 19/748, 20 November 1746.
33. Ms GAA Ondertrouwregister 706/185, 30 March 1708.
34. Oldewelt, *Kohier*, Koeman, *Atlantes Neerlandici*, 4, pp.222 & 410 and Gernez, 'Le libraire neerlandais'.
35. Clifton, *Directory*. Harris, *A Treatise*. Taylor, *Mathematical Practitioners 1714-1840*.
36. Brewington, 'Notes on the cross-staff', p.191.
37. Van der Krogt, *Advertenties*, no.1430. Ms NSA 'Alphabetical list', 1986.
38. Cox, 'De jacobsstaf', p.47. Ms NSA 'Alphabetical list', 1986.
39. *Catalogus der voorwerpen*.
40. *Catalogus der belangrijke bibliotheken* and *Byzondere veiling*.
41. Ms GAA Poortersboek 28/171, 8 December 1785.
42. Ms GAA Poortersboek 19/426, 24 mei 1742.
43. Ms GAA Ondertrouwregister 686/335, 24 April 1666.
44. Davids, 'Van Lastman naar Gietermaker', p.162.
45. Ms GAA Ondertrouwregister 616/492, 3 May 1771.

46. Ms GAA Poortersboek 25/174, 13 February 1772, his father-in-law was a printer.
47. De Jonge, *De opkomst*, p.183. Ms NSA 'Alphabetical list', 1986.
48. Ms GAA Poortersboek 15/88, 1 June 1723 & 21/420, 3 September 1750.
49. Ms GAA Ondertrouwregister 508/382, 3 May 1681.
50. Ms GAA Ondertrouwregister 479/388, 13 February 1659.
51. Ms GAA 'Kohier 200e penning, p.213, wijk 32'.
52. Koeman, *Atlantes Neerlandici*, 4, pp.276-279. Ms NSA 'Alphabetical list', 1986.
53. Kok, 'Cartografie', p.24. Ms NSA 'Alphabetical list', 1986.
54. In Jean van Keulen, *Le grand Nouveau Atlas de la Mer*. Amsterdam, 1680.
55. Mörzer Bruyns, 'De instrumenten', p.62.
56. Kok, 'Cartografie', p.36. Ms NSA 'Alphabetical list', 1986.
57. Mörzer Bruyns, 'De instrumenten', p.64.
58. Kok, 'Cartografie', pp.37-38.
59. Ms GAA Poortersboek 21/381, 24 January 1753 & 27/364, 1 June 1780.
60. Ms GAA Ondertrouwregister 596/86, 29 September 1752.
61. Bedini, 'At the Sign of the Compass'.
62. Ms GAA Ondertrouwregister 697/156, 1 June 1691.
63. Koeman, *Atlantes Neerlandici*, 4, pp.409-411. Ms NSA 'Alphabetical list', 1986.
64. Gernez, 'Le libraire neerlandais', p.67.
65. Van der Krogt, *Advertenties*, no.983.
66. Koeman, *Atlantes Neerlandici*, 4, pp.223-226. Ms NSA 'Alphabetical list', 1986.
67. Ms GAA Poortersboek 19/748, 20 September 1746.
68. Ms GAA Poortersboek 19/36, 18 May 1747.
69. Ms GAA Ondertrouwregister 590/203, 14 April 1747 & 605/265, 14 August 1761.
70. Ms GAA Poortersboek 21/267, 11 December 1750, his father was a glazier.
71. Koeman, *Atlantes Neerlandici*, 4, pp.223-224. Rooseboom, *Bijdrage*, p.101 and Gernez, 'Le libraire neerlandais'. Ms NSA 'Alphabetical list', 1986.
72. Van Nierop, 'Naamregister'.
73. Ms GAA Poortersboek 27/26, 17 January 1782, his father, Dick Jansz. Noordijk, was a compass maker.
74. Ms GAA Ondertrouwregister 750/253, 8 May 1778 & 635/62, 9 May 1790.
75. Bos, 'Vragenrubriek'. Ms NSA 'Alphabetical list', 1986.
76. Clifton, *Directory*. Bryden, *Scottish Scientific Instrument-Makers*, pp.4-5 & 55. Taylor, *Mathematical Practitioners*, p.277.
77. Ms GAA Poortersboek 8/636, 7 December 1695 & 11/282, 6 April 1707.
78. Ms GAA Ondertrouwregister 494/404, 5 April 1670.
79. Koeman, *Atlantes Neerlandici*, 4, pp.436-437. Ms NSA 'Alphabetical list', 1986.
80. Ms GAA Poortersboek 17/350, 15 May 1732, his father, Gerrit Lucas van Schagen, was engraver.
81. Ms GAA Poortersboek 19/406, 21 July 1747.
82. Clifton, *Directory*. Taylor, *Mathematical Practitioners*, pp. 244-245, 295, 380, 383 & 402.
83. Ms GAA Ondertrouwregister 494/445, 18 April 1670.
84. Ms GAA Ondertrouwregister 692/109, 10 August 1680 & Notarieel Archief 4495, act. 229, 9 August 1680.
85. Ms GAA Ondertrouwregister 540/319, 12 November 1706.
86. Clifton, *Directory*. Taylor, *Mathematical Practitioners*, pp.412 & 416.
87. Ms GAA Ondertrouwregister 492/32, 21 April 1668.
88. Ms GAA Poortersboek 15/46, 24 October 1725 & 19/667, 20 June 1747.
89. Ms GAA Poortersboek 17/275, 1 September 1735 & 24/287, 1765. His father, Jan Vouwens, was wine merchant.
90. Clifton, *Directory*. Taylor, *Mathematical Practitioners*, pp.213-215, 350 & 373. Wakely, *The Mariner's Compass Rectified*.
91. Oldewelt, *Kohier*. Ms NSA 'Alphabetical list', 1986.
92. Davids, *Zeewezen en wetenschap*, p.121.
93. Clifton, *Directory*.

Appendix II

1. Paid for a cross-staff given to Martin Frobisher in 1576. Waters, *The Art of Navigation*, p.530.
2. Davids, *Zeewezen en wetenschap*, p.121.
3. Ms ARA VOC 103.
4. Ms ARA VOC 5017.
5. Ms ARA VOC 5018.
6. Ms ARA VOC 7150. See Appendix III.
7. Ms ARA VOC 119.
8. Ms ARA VOC 7407.
9. Ms NSA 'Kasboek', 27/12/1759.
10. Ms ARA VOC 5023.
11. Ms NSA 'Kasboek', 18/12/1765.
12. In a sales catalogue of instruments made and sold by the firm of Van Keulen, in a copy of Klaas de Vries, *Schat-kamer ofte Konst der Stierlieden*, Amsterdam, 1777, in the library of the NSA.
13. Ms ARA VOC 5018.
14. Ms ARA VOC 5018.

Appendix III

1. I am grateful to Professor Dr Günther Schilder of Utrecht University for providing the information for this appendix.
2. Ms ARA VOC 7150.
3. 127 guilders and 10 *stuivers* (20 *stuivers* to the guilder).
4. Ms ARA VOC 7151.
5. With scales for altitude and zenith distance.
6. Brass apertures.
7. Ms ARA VOC 7152.
8. Ms ARA VOC 7153.
9. Ms ARA VOC 7154.
10. Ms ARA VOC 7155.

Appendix V

1. For the conversion into mm, use was made of J.H. Alexander, *Universal Dictionary of Weights and Measures, Ancient and Modern reduced to the standards of the United States of America*. Baltimore, 1850.

Index of names of persons and ships

(ships' names in italics)

General index